The Silence of Slime Mould
Art works by Andrew Adamatzky

Andrew Adamatzky

The Silence of Slime Mould Art Works

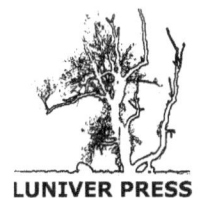

LUNIVER PRESS

Published by Luniver Press Bristol BS39 5RX United Kingdom

British Library Cataloguing-in-Publication Data
A catalogue record for this book is available from the British Library

The Silence of Slime Mould. 2015

Copyright ©Andrew Adamatzky 2015

All rights reserved. This book, or parts thereof, may not be reproduced in any form or by any means, electronic or mechanical, including photocopying, recording or by any information storage and retrieval system, without permission in writing from the copyright holder.

ISBN-10: 1-905986-42-4
ISBN-13: 978-1-905986-42-2

While every attempt is made to ensure that the information in this publication is correct, no liability can be accepted by the authors or publishers for loss, damage or injury caused by any errors in, or omission from, the information given.

To those who made my life good

Acknowledgements

The art works are produced at nights, while I was taking breaks from leading the Leverhulme Trust research project "Mould intelligence: biological amorphous robots", the Samsung project "A feasibility study on interfacing nano-, and macro-worlds via amorphous biological computing substrate" and the European project FP7-ICT-2011-8 "Physarum Chip: Growing computers from slime mould". I am thankful to Theresa Schubert (Bauhaus-Universität, Weimar) for inspiring me to convert my scientific results to art works, and to Dee Smart (UWE, Bristol) for helping me stage my first exhibition "Slime Mould Machines" at At-Bristol Science Centre, to Nicoletta Gori (London) for photo documenting the exhibition, to Prof. Glenn Lyons and Prof. Paul Olomolaiye (UWE, Bristol) for supporting all my activities.

Contents

About author	1
Why slime mould?	3
Germination	6
Maze	7
Wrapping I	8
Wrapping II	9
Encapsulation	10
Choice	11
Physarum chip	12
Slime wires I	13
Delays	14
Geometrically constrained	15
Slime roads	16
Discs	17
Slime wires II	18
Highways	19
Milky Way	20
Cartwheel galaxy	21
Slimy Autobahns	22
Nutrient rich substrate	23
Slime eye	24
Voronoi diagram	25
Innervation I	26
Innervation II	27
Highways II	28

Slime wires III 29
Balkans 3D 30
Exploration 31
World colonisation I. The Silk Road 32
Galaxy II 33
World colonisation II 34
Route planning 35
Valerian versus Kalms Sleep 36
Slime whiskers 37
Colonisation of Mars 38
Concave hull 39
Active zone 40
Nutrient substrate II 41
Propagation 42
Maze II 43
Eye 44
Ice England 45
The growth 46
Excitation waves 47
Voronoi diagram II 48
Plant based one bit full binary adder 49
Proximity graph 50
References 53
Index 55

About author

Andrew Adamatzky is a scientist, artist and philosopher living in Bristol, United Kingdom. He is a professor in unconventional computing, UWE, Bristol. His research interests span computer science, biology, mathematics, physics and engineering. Adamatzky excels in the fields of reaction-diffusion computing, cellular automata, physarum computing, massive parallel computation, applied mathematics, collective intelligence and robotics. He authored several high-profile books, including "Dynamics of Crowd Minds" (2005), "Physarum Machines" (2010) and "Reaction-Diffusion Automata" (2013).

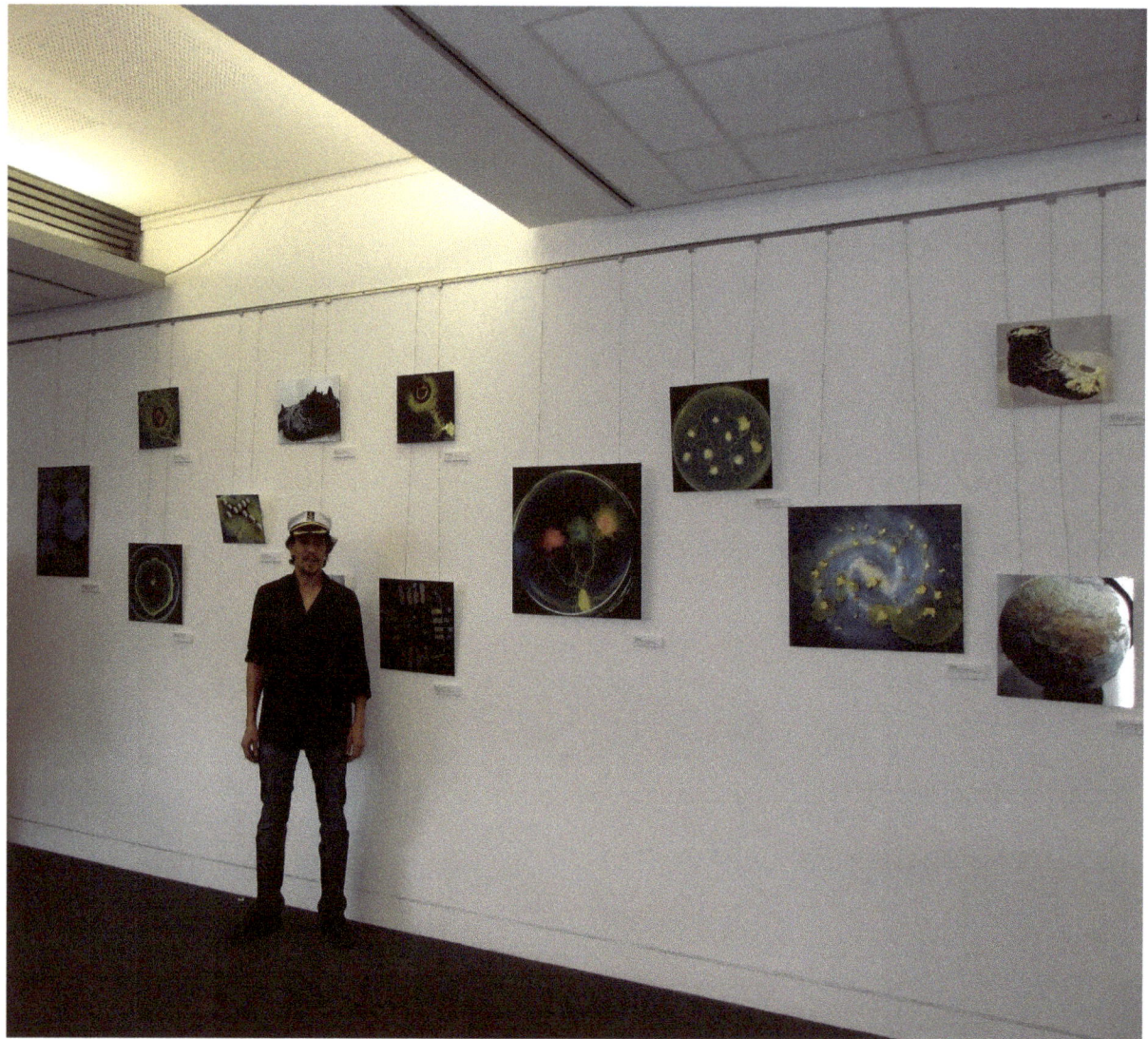

Andrew Adamatzky. Slime Mould Machines exhibition at Bristol Bright Night, 2014

Why slime mould?

I see the world through the patterns generated by spatially extended active non-linear substrates [1, 20, 9]. The artistic works present my vision of massive-parallel sensing and perception of the world according to biological, chemical and physical systems. By producing my art works, I develop an intuition of how distributed amorphous concurrent and decentralised sensorial fusion, information processing and decision making are implemented in natural systems.

Most of the art works in this book relate to slime mould. The plasmodium of acellular slime mould *Physarum polycephalum* is a gigantic single-cell organism visible by the unaided eye [24]. *P. polycephalum* consumes microscopic particles, and as it conducts its foraging behaviour, the plasmodium spans scattered sources of nutrients with a network of protoplasmic tubes. The cell shows a rich spectrum of behavioural morphological patterns in response to the changing environmental conditions. The plasmodium optimises its protoplasmic network that covers all sources of nutrients and guarantees robust and quick distribution of nutrients in the plasmodiums body. Given the data represented by chemical or physical stimuli, we can employ and modify the behaviour of the slime mould to direct it to solve a range of computing and sensing tasks [5]. Plasmodium's foraging behaviour can be interpreted as a computation; data are represented by the spatial characteristics of attractants and repellents, and the results are represented by the structure of the protoplasmic network. Plasmodium can solve computational problems with natural parallelism; e.g. related to the shortest path, hierarchies of planar proximity graphs, computation of plane tessellations, execution of logical computing schemes, and natural implementation of spatial logic and process algebra.

Through my art works with the slime mould *P. polycephalum*, I show that the absence of a brain, or even a nervous system, does not absolve a living creature from intelligence. Slime mould solves hard geometrical and optimisation problems without any nervous system, human-like logic or reasoning.

Living substrates compute by changing their shape, activity or location [1]. Physarum machines are computers made from a huge living cell, the plasmodium of acellular slime mould *P. polycephalum*. They are amorphous biological substrates which transform data, represented by gradients of attractants and repellents, into results of computation, represented by the morphology of the slime mould [5].

Slime mould is attracted to and eats oat flakes; it smells the oat flakes from far away. This is why you often see oat flakes in the pictures. They are essential tools of programming the slime mould computers.

The slime mould is a robust creature. It can propagate on a variety of substrates, from aluminium foil to plastic — seen in the works *Germination* and *Exploration*. When substrate is favoured by the slime mould (e.g. a nutrient-rich agar gel), the slime mould propagates as a wave in all directions. The works *Nutrient rich substrate*, *Nutrient substrate II* and *Propagation* illustrate the slime moulds propagation in comfortable conditions. These patterns of the slime mould propagation are phenomenologically similar to the excitation waves in active non-linear media (e.g. Belousov-Zhabotinsky excitable chemical systems [20]: the work *Excitation waves*). The slime mould waves can solve a range of computational tasks. A plane tessellation, or Voronoi diagram, is the most famous problem, seen in works *Voronoi diagram* and *Voronoi diagram II*. We can control the propagation of the slime mould waves by tuning the shapes of their substrates [13]: *Delays* and *Geometrically constrained*. The slime mould speeds up in the narrow places but slows down in the widening — *Discs*.

In harsh environmental conditions, the slime mould grows like a tree branch — *Active zone*. The slime mould computes with its growing branches [22]. When spanning several sources of nutrients, the slime mould develops protoplasmic networks resembling vascular networks — *The Eye*. These networks are akin to mathematical structures known as proximity graphs [2]: *Proximity graph*. The slime mould propagating on a non-nutrient substrate is eager to find sources of food. By following gradients of chemo-attractants, the slime mould grows along the shortest path from its original inoculation site to the closest source of food [11]. The works *Maze* and *Maze II* demonstrate that the slime mould can solve a labyrinth problem. I put a fertile oat flake in a central chamber of a plastic maze and place a piece of the slime mould in an outside chamber. Chemo-attractants were diffused from the fertile oat flake. The slime mould followed the gradients of the chemoattractants. Thus, the slime mould found the path towards the central chamber [11].

Networks of protoplasmic tubes spanning sources of nutrients resemble human-made road networks [8]. To validate how well the slime mould can approximate the road networks, I placed oat flakes in major urban areas of a country and inoculated the slime mould at the capital of the country. The slime mould transport networks satisfactorily matched the motorways, highways and Autobahns, seen in the works *Highways*, *Highways II*, *Slime roads*, *Route planning* and *Slimy Autobahns*. Experiments on a globe and three-dimensional models uncovered the slime moulds abilities to imitate large-scale historical developments, as demonstrated in the formation of *The Silk Road*, and emergence of civilisation in *Balkans* [12]: the works *World colonisation I: The Silk Road*, *World colonisation II* and *Balkans 3D*. The experimental techniques developed have also been applied to imitate the growth of galaxies and space exploration [14] — *Milky Way*, *Cartwheel galaxy* and *Galaxy II* — and the colonisation of planets: *Colonisation of Mars*.

The works *Wrapping I*, *Wrapping II*, *Encapsulation* and *Concave hull* emerged from the experimental studies of the slime moulds unusual attraction to pills containing extracts of somniferous plants [10]. The slime mould propagates towards the pills and encloses them into a fine network of protoplasmic tubes. However, the slime mould never physically touches a pill. This finding helped me to design the slime mould computer, which approximates hulls of planar data sets [11]. I later confirmed that the slime mould is attracted to valerian roots [6]. The work *Valerian versus Kalms Sleep* shows a typical experiment on binary preferences of the slime mould to dry plant samples [6].

The art work *Choice* shows the slime mould propagating from the edge of a Petri dish towards oat flakes saturated in a solution of food colouring. This work was a byproduct of my research on the transport of substances by the slime mould [4]: the slime mould intakes a substance, distributes the substance inside its body and then deposits the substance somewhere far away from the original site of intake [4].

During one of our European projects, we aimed to build functional biomorphic computing devices operated by the slime mould *P. polycephalum* [21]. We found that — combined with conventional electronic components in a hybrid chip —- the slime networks improved self-repairing features [14] and added some unique sensing properties to digital and analog circuits [23, 15]. The works *Physarum chip*, *Slime wires I* and *Slime wires II* illustrate designs of the slime mould chips and their interface with conventional electronics.

The slime mould is capable of sensing tactile, chemical and optical stimuli and converting the stimuli to characteristic patterns of its electrical potential oscillations [17, 26, 16, 25]. The electrical responses to stimuli may propagate along protoplasmic tubes for distances exceeding tens of centimetres, like impulses in neural pathways do. A slime mould makes decision about its growth direction based on information fusion from thousands of spatially extended protoplasmic loci, similarly to a neuron collecting information from its dendritic tree. The analogy between the nervous system and the *P. polycephalum* is distant yet inspiring. We can speculate whether alternative (would-be) nervous systems can be developed and practically implemented from the slime mould [19]. Four of my works related to a would-be nervous system made of the slime [19] are *Slime eye*, *Innervation I*, *Innervation II* and *Slime whiskers*. The work *Slime eye* shows an experiment on the implementation of optical pathways with the slime mould. A sensorial innervation of the front scalp made of the slime mould is illustrated in the works *Innervation I* and *Innervation II*. A distinct electrical response from the slime mould to tactile stimulation is employed in the design of the slime mould-based whiskers [17], as illustrated by the work *Slime whiskers*.

What will be after the slime mould computers exhaust their potential? The future non-slime-mould computers will employ spatio-temporal dynamics of crystallisation [3, 7] — *Ice England* and *The growth* — and, electrical properties of plant roots [18] — *Plant based one bit full binary adder*.

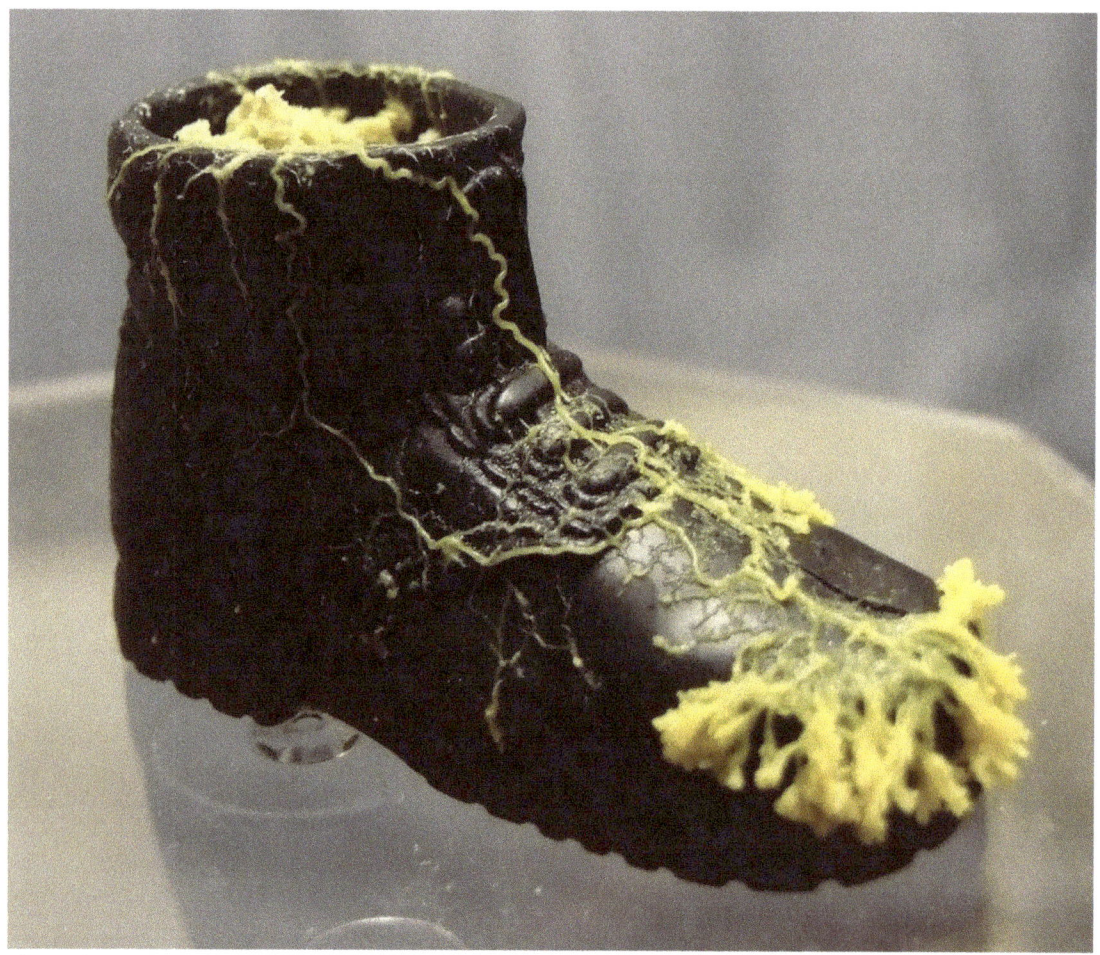

Germination.
35 × 28 cm. Direct printing on brush-finished aluminium.

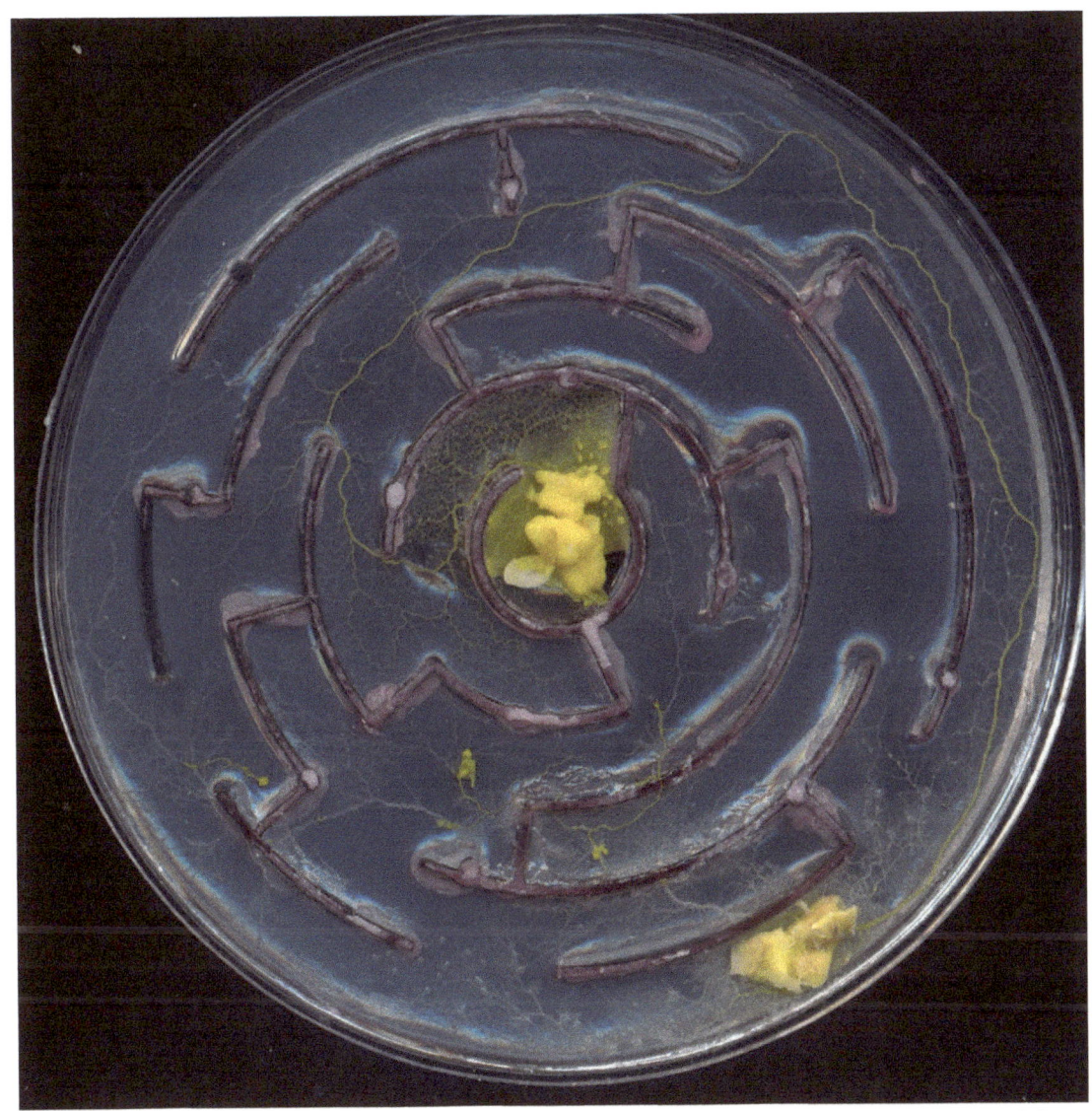

Maze.
30×30 cm. Direct printing on brush-finished aluminium.

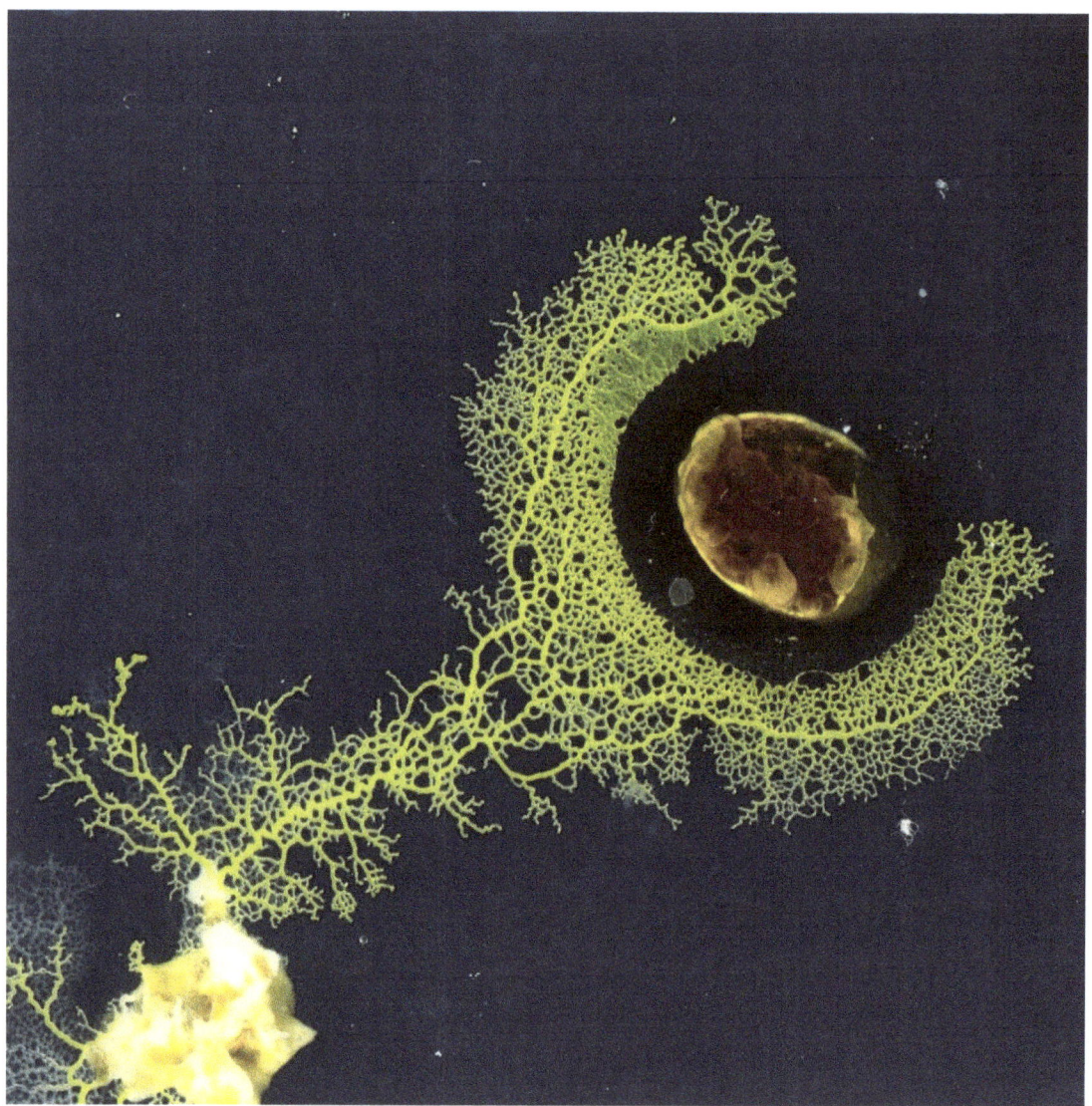

Wrapping I.
30 × 30 cm. Direct printing on brush-finished aluminium.

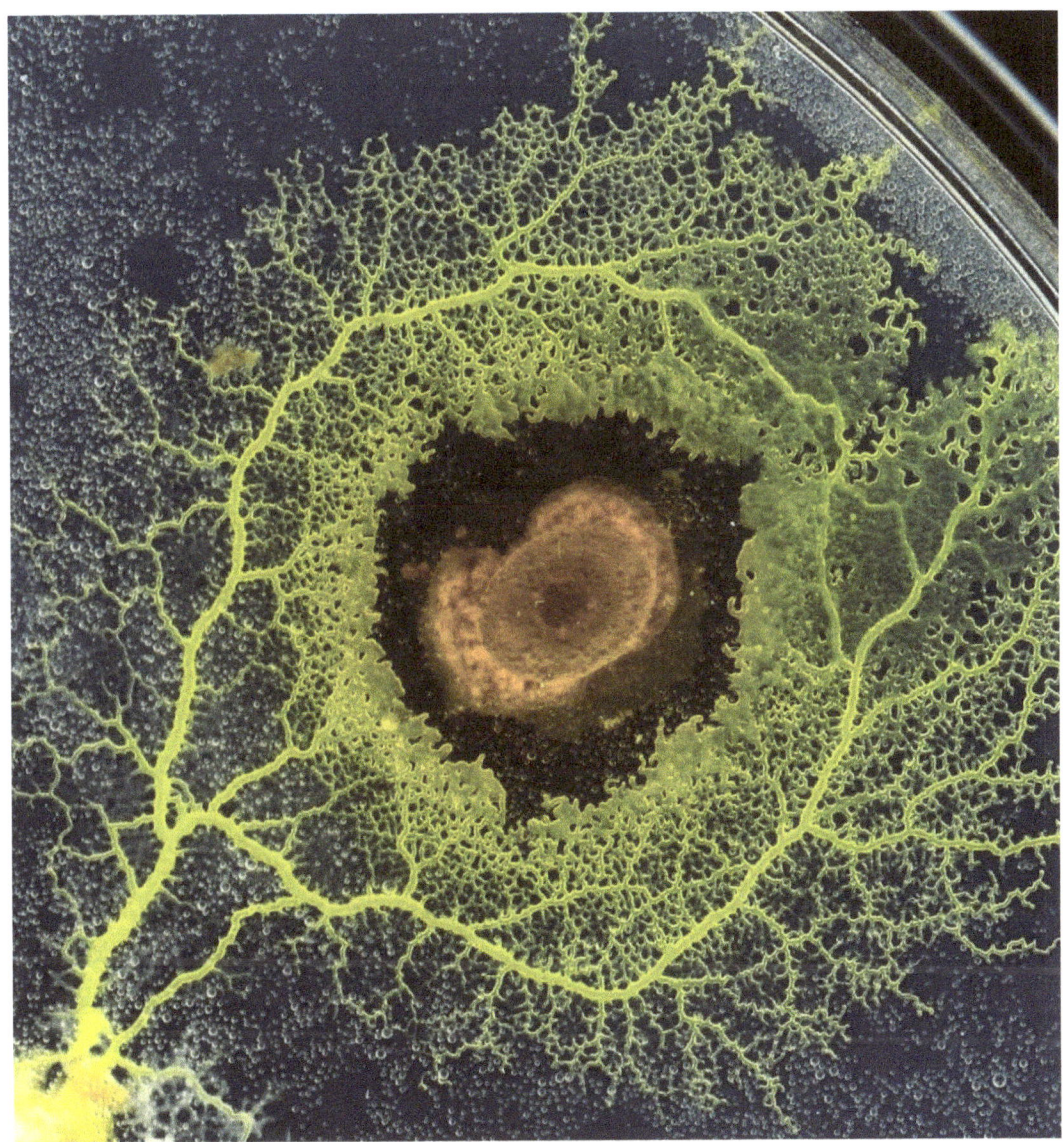

Wrapping II.
30 × 30 cm. Direct printing on brush-finished aluminium.

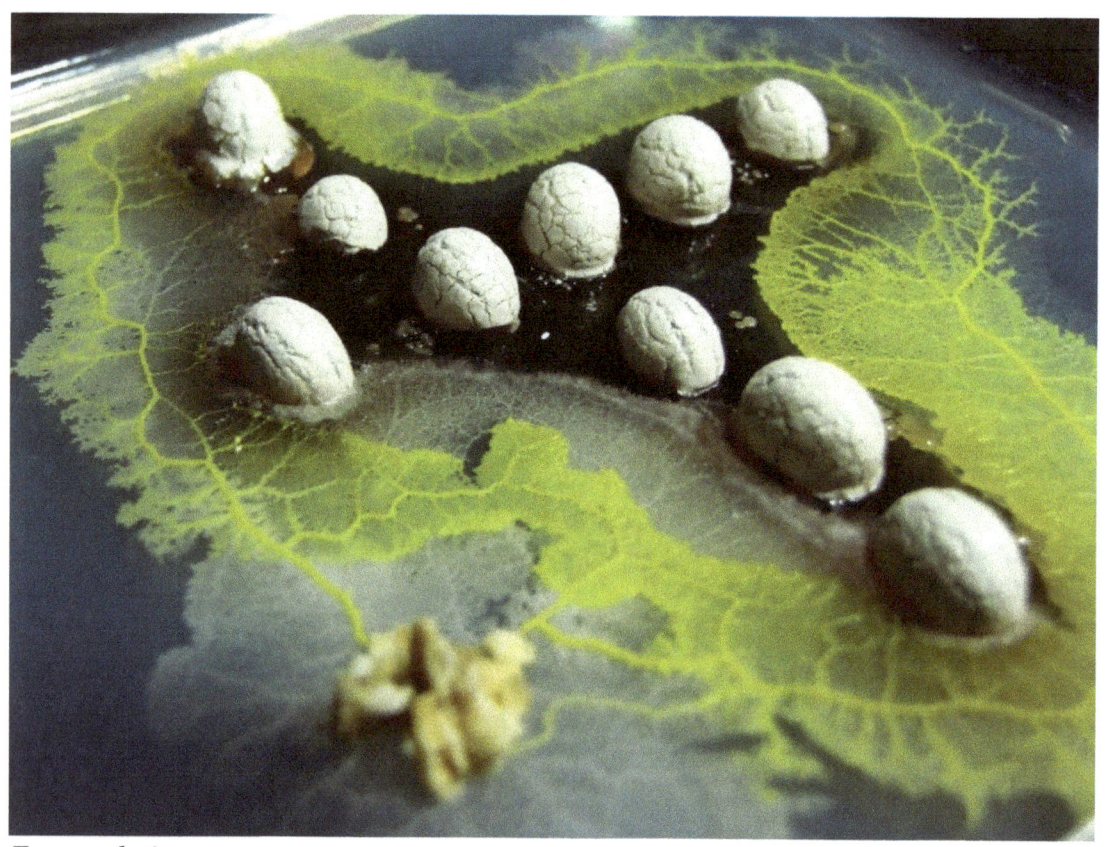

Encapsulation.
30 × 24 cm. Direct printing on brush-finished aluminium.

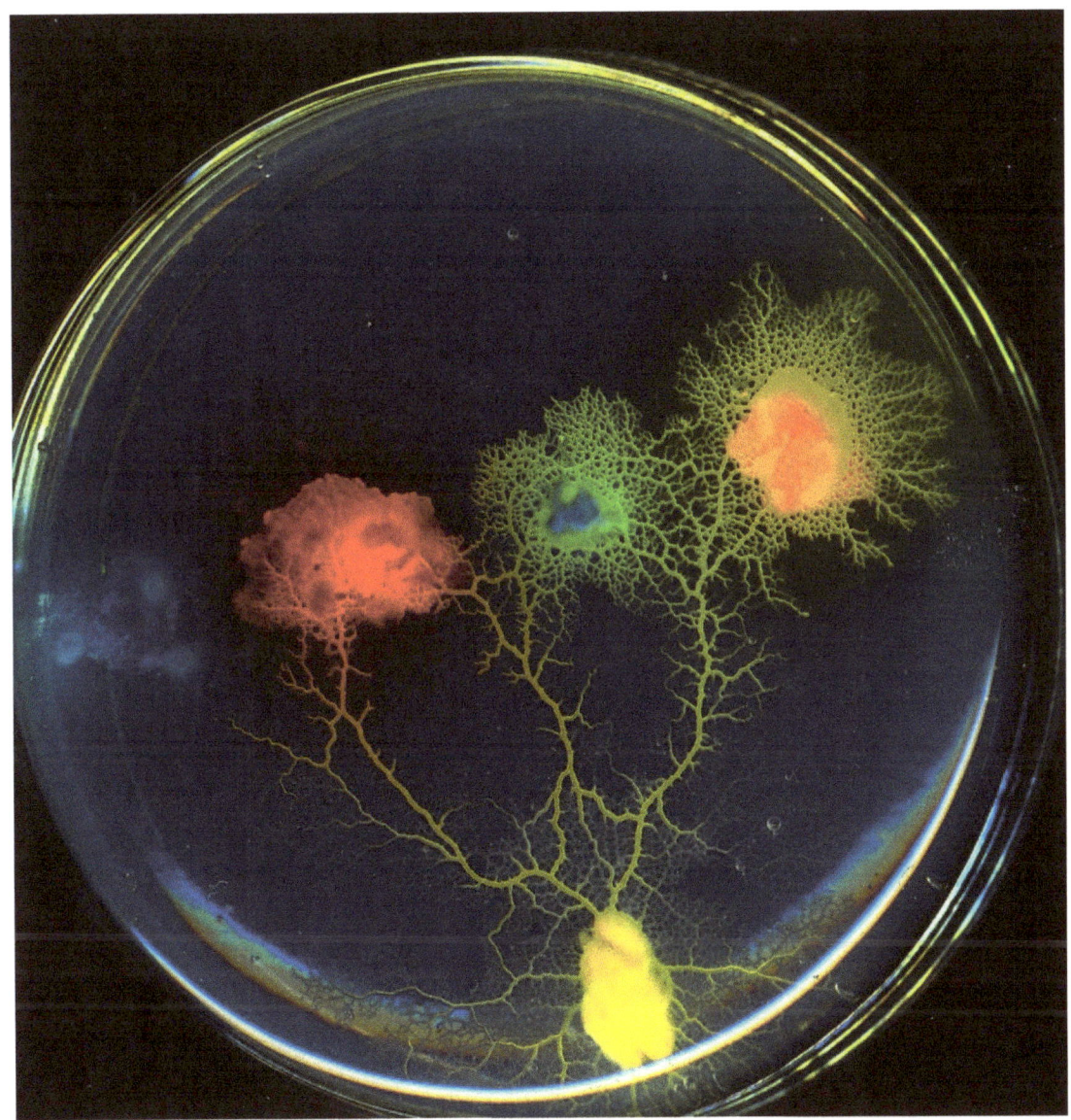

Choice.
60×60 cm. Direct printing on brush-finished aluminium.

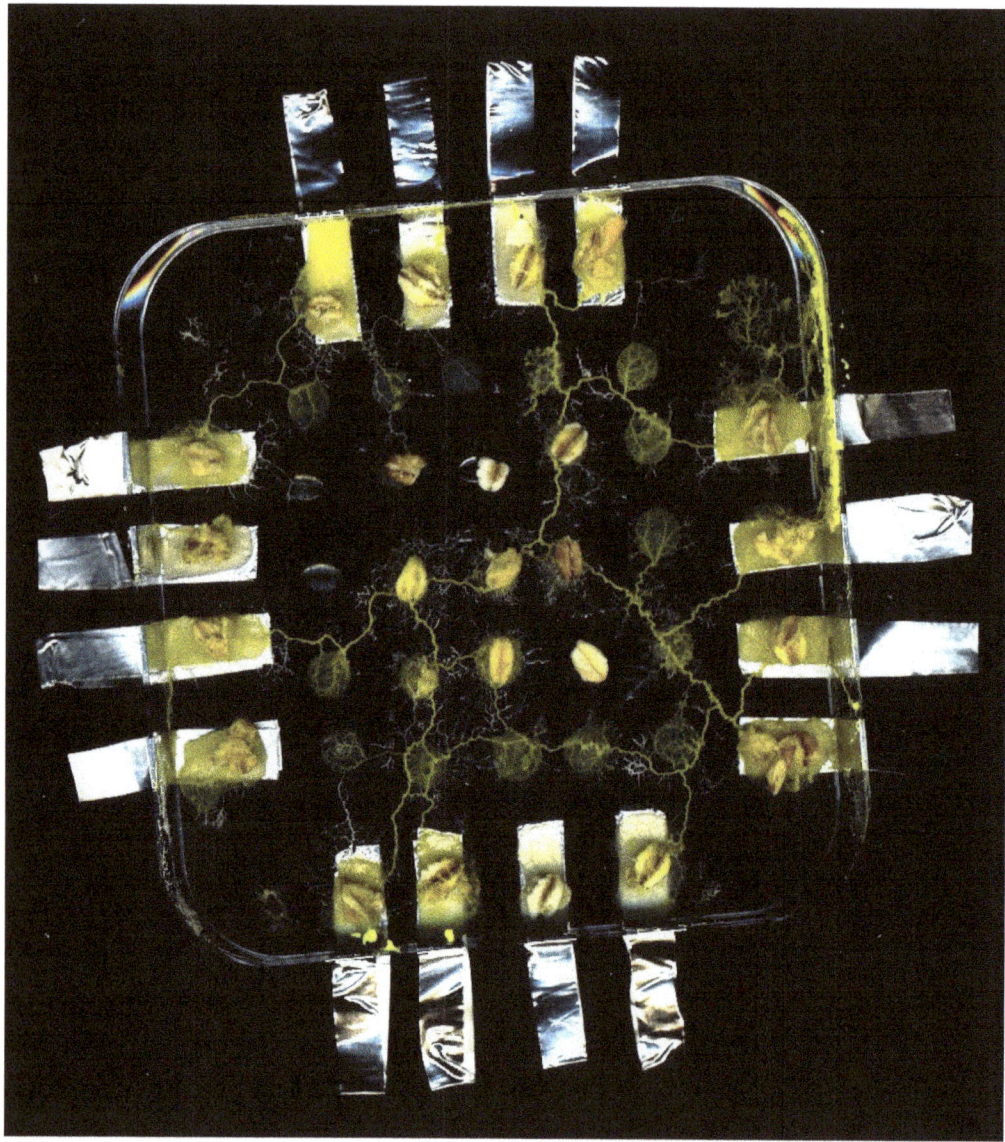

Physarum chip.
40×40 cm. Direct printing on brush-finished aluminium.

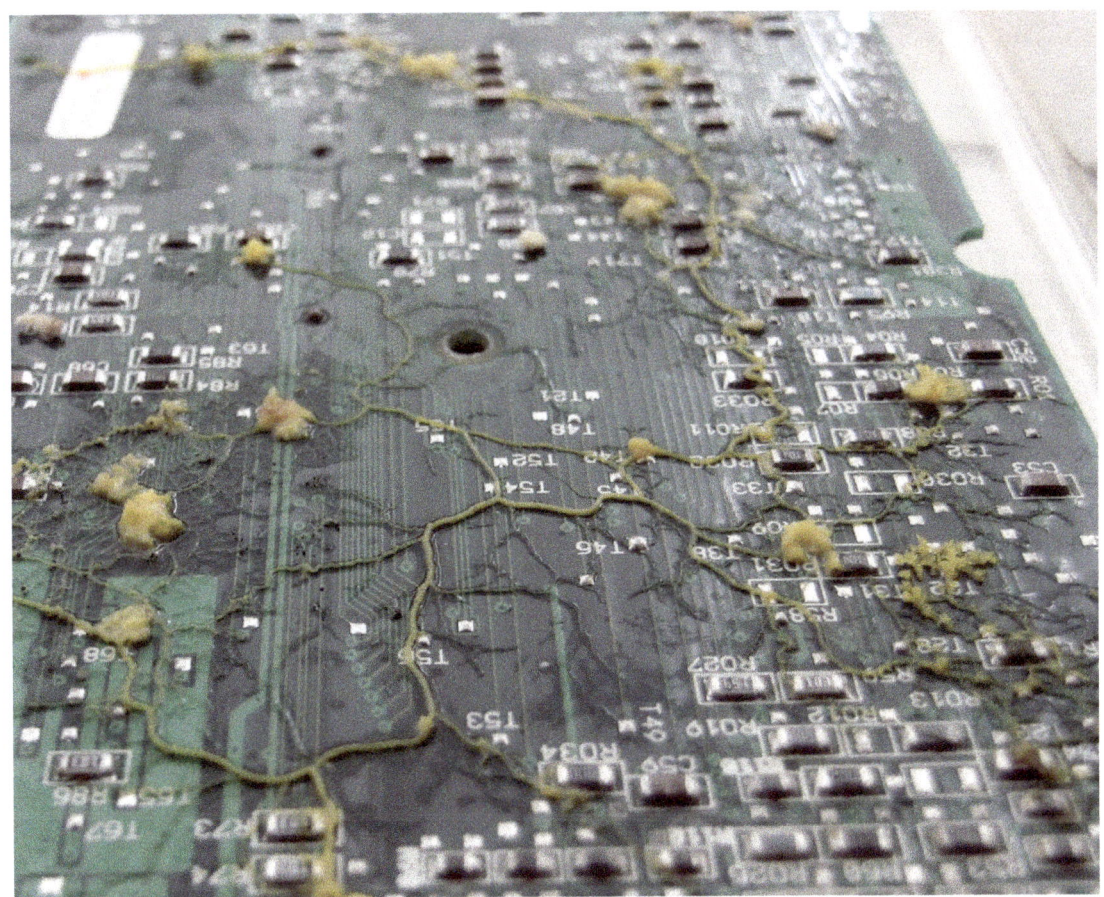

Slime wires I.
35×28 cm. Direct printing on brush-finished aluminium.

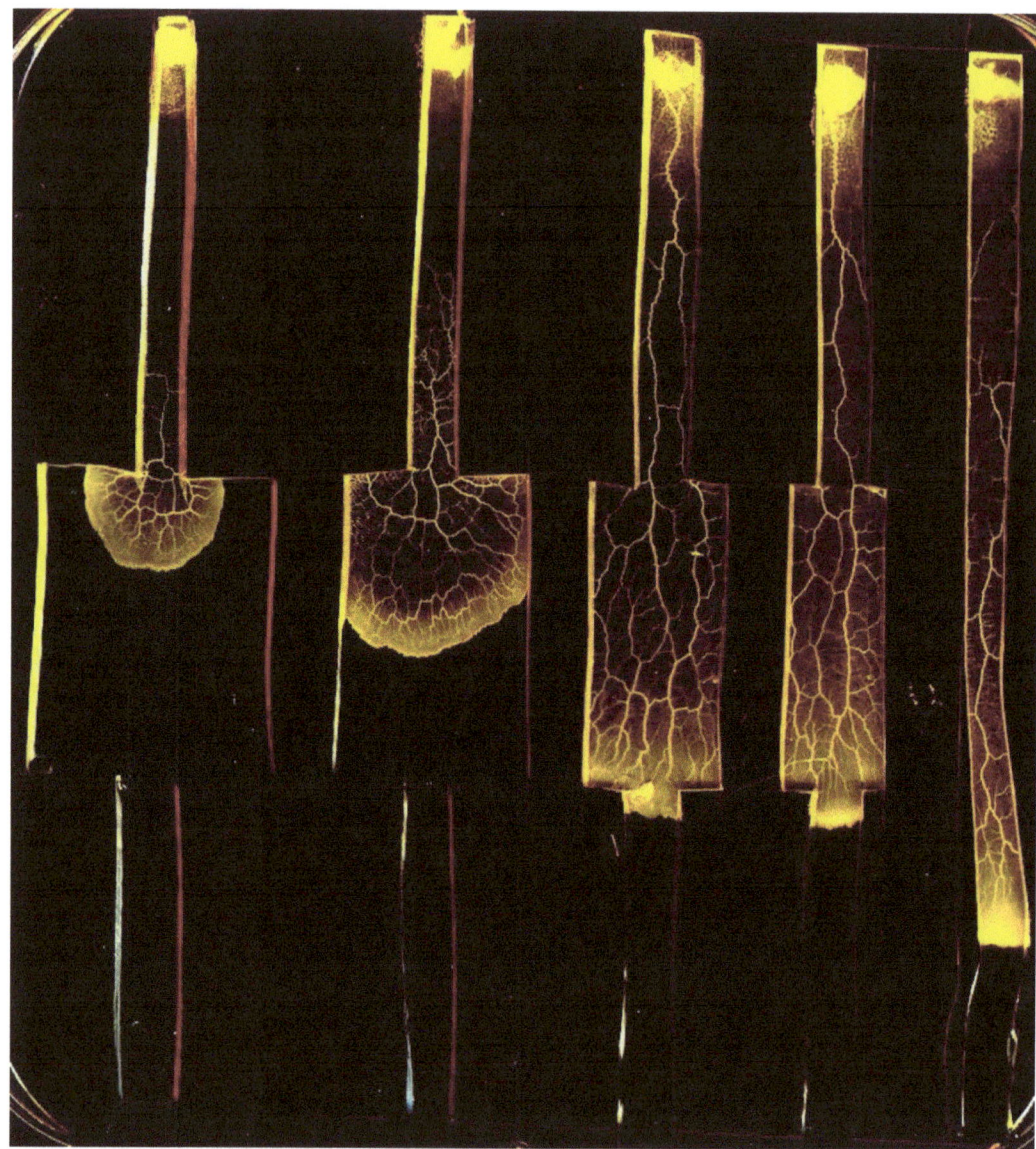

Delays.
20×20 cm. Direct printing on brush-finished aluminium.

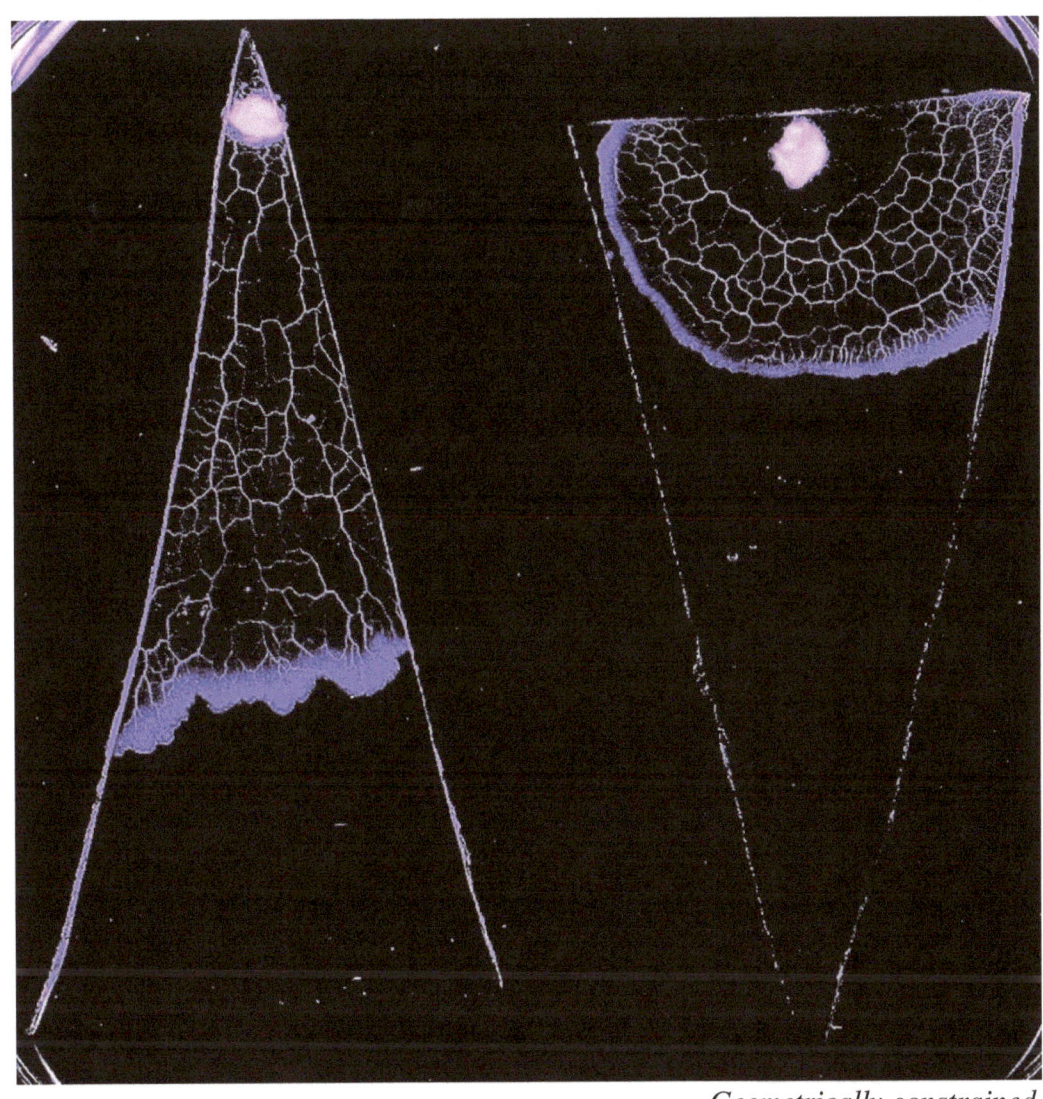

Geometrically constrained.
20×20 cm. Direct printing on brush-finished aluminium.

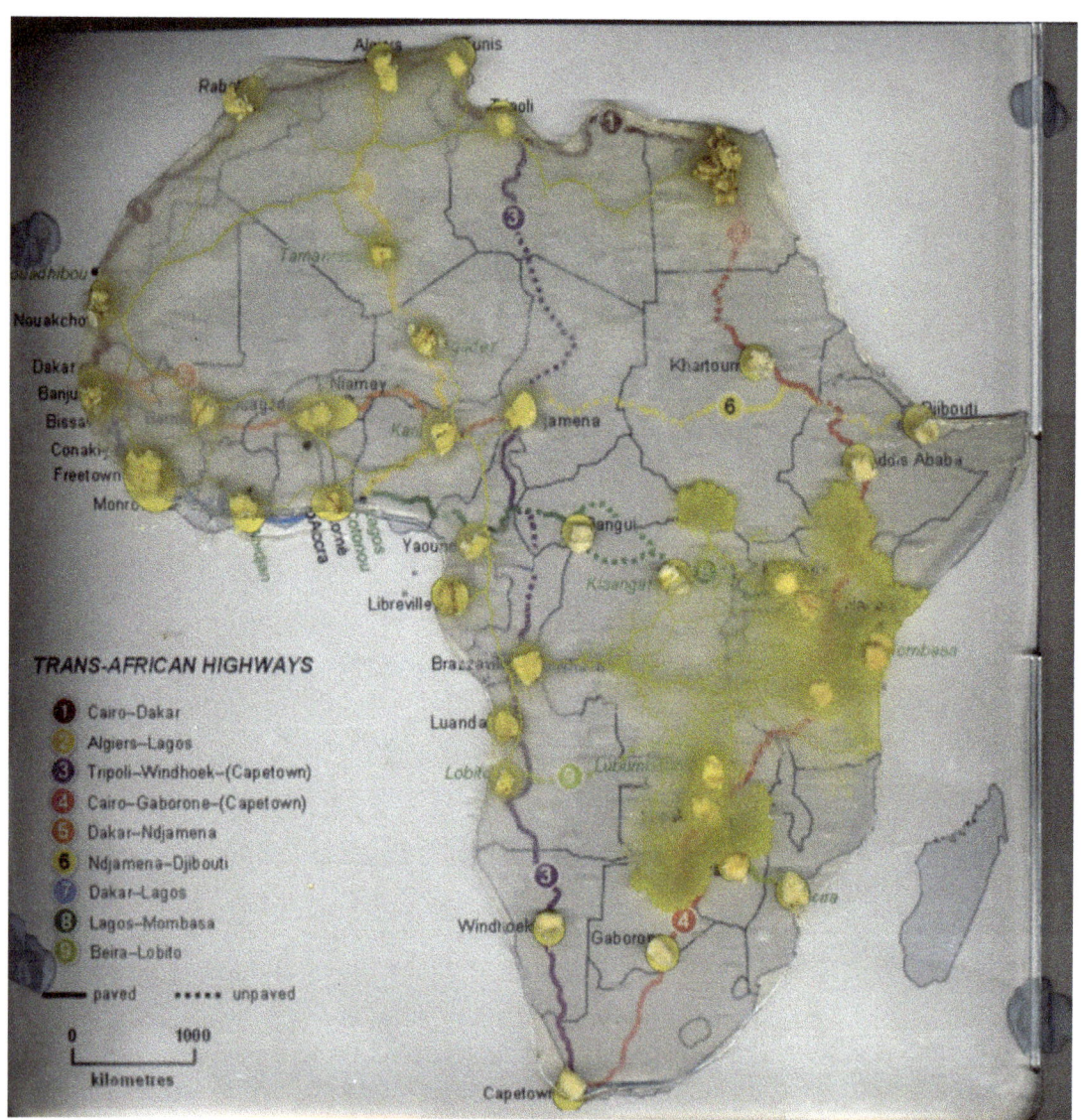

Slime roads.
40×40 cm. Direct printing on brush-finished aluminium.

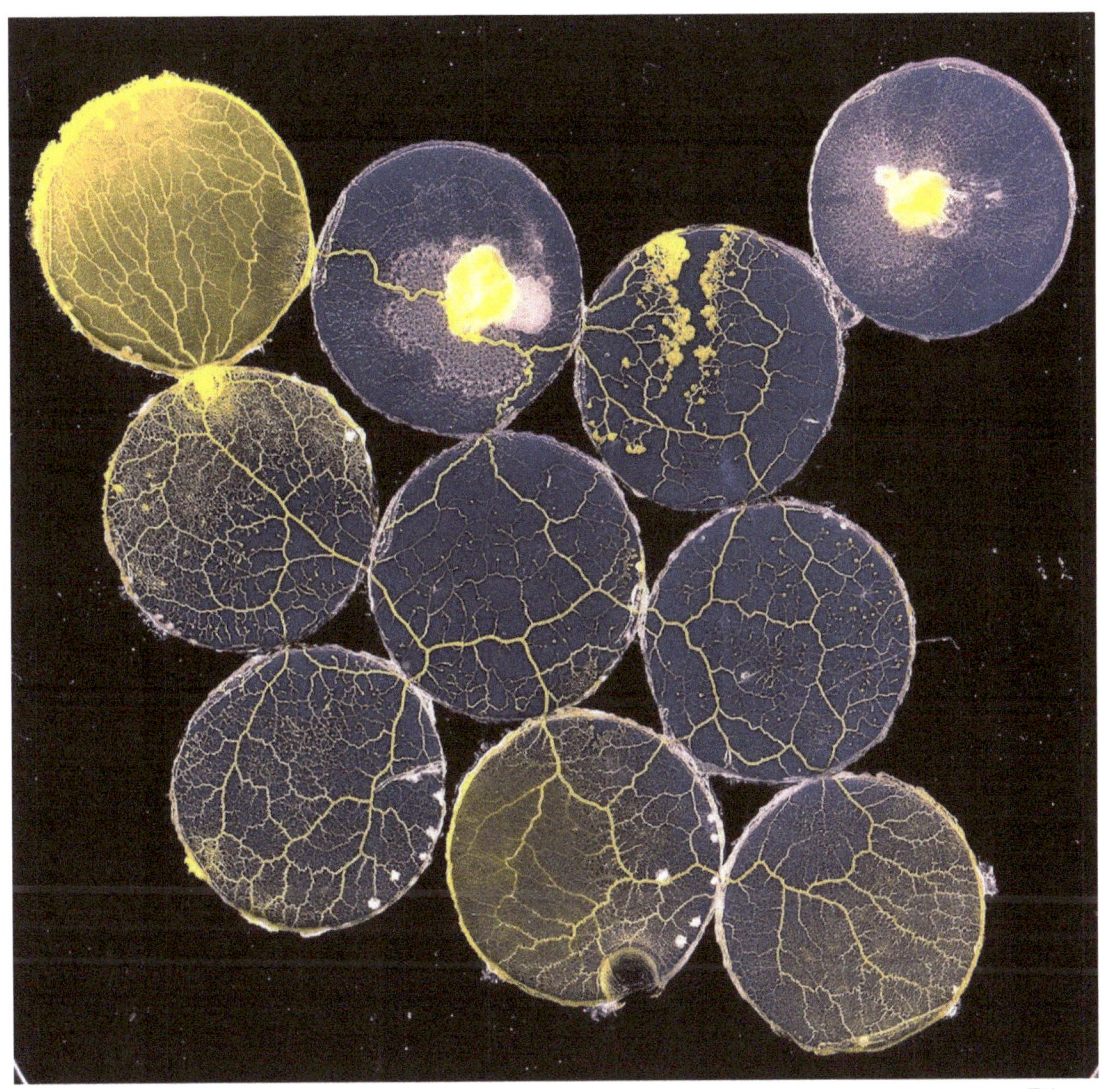

Discs.
50×50 cm. Direct printing on brush-finished aluminium.

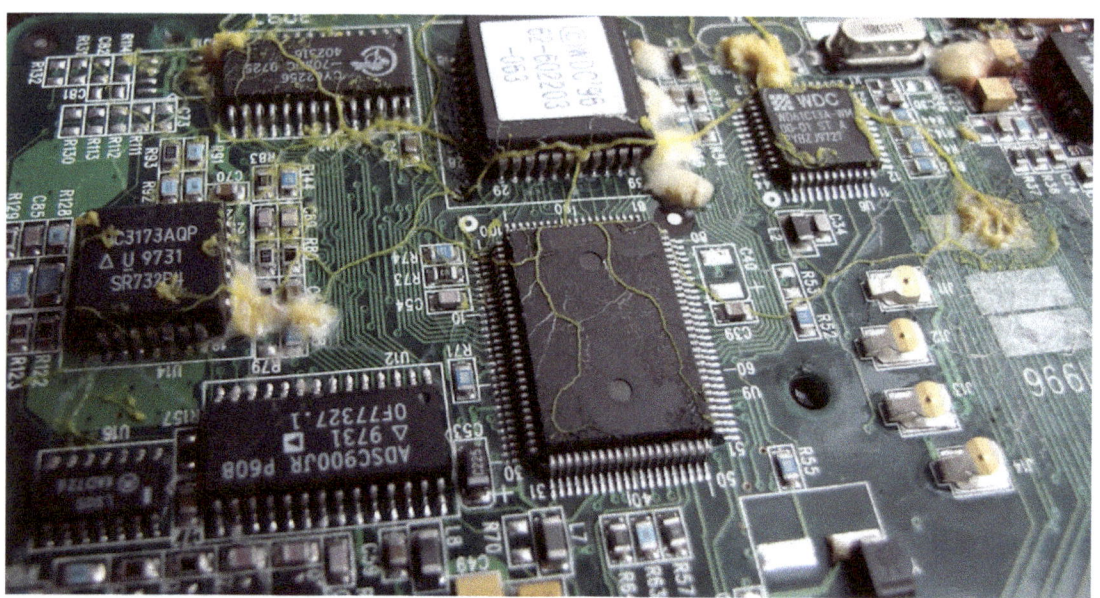

Slime wires II.
48×27 cm. Direct printing on brush-finished aluminium.

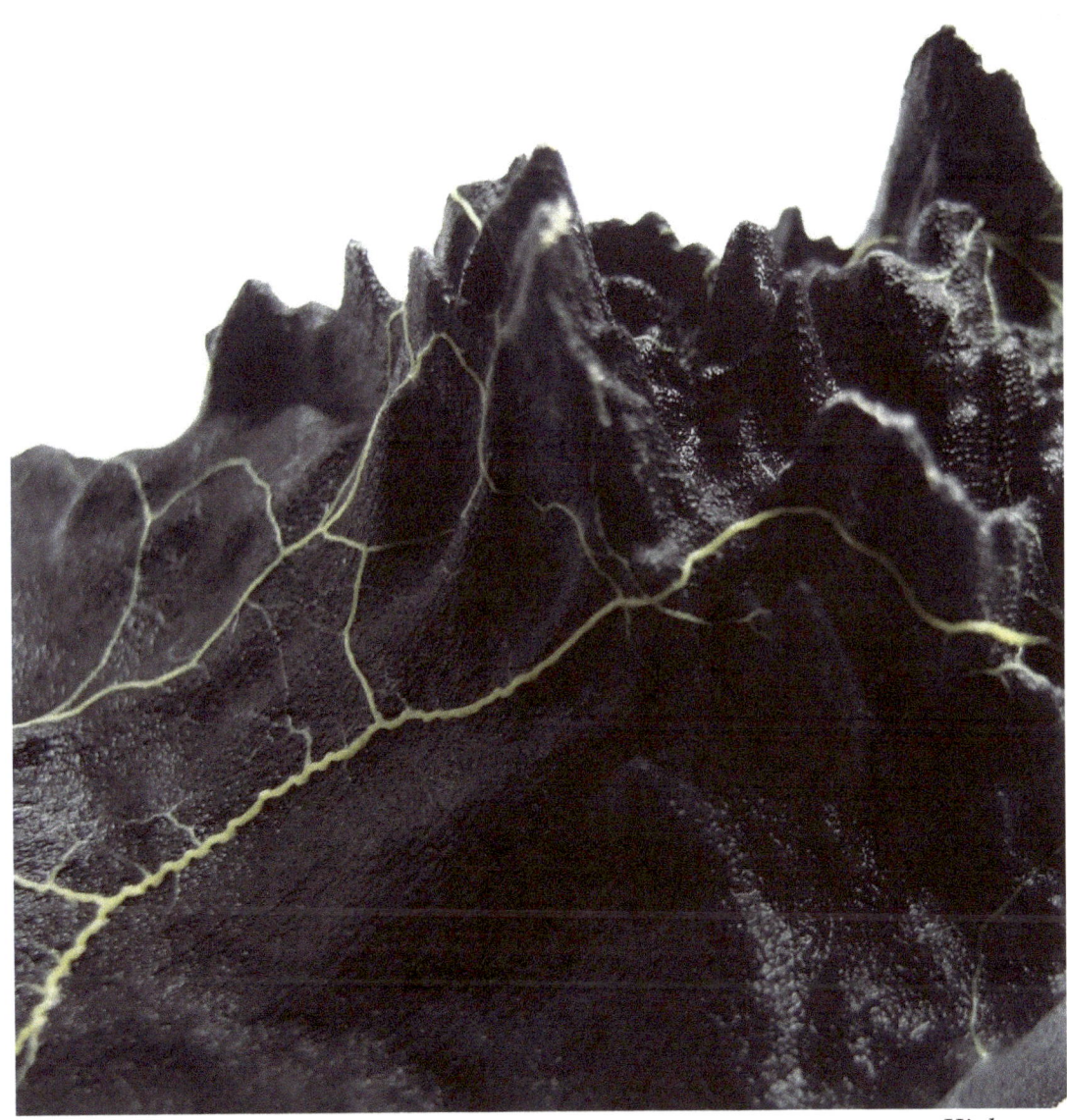

Highways.
40×40 cm. Direct printing on brush-finished aluminium.

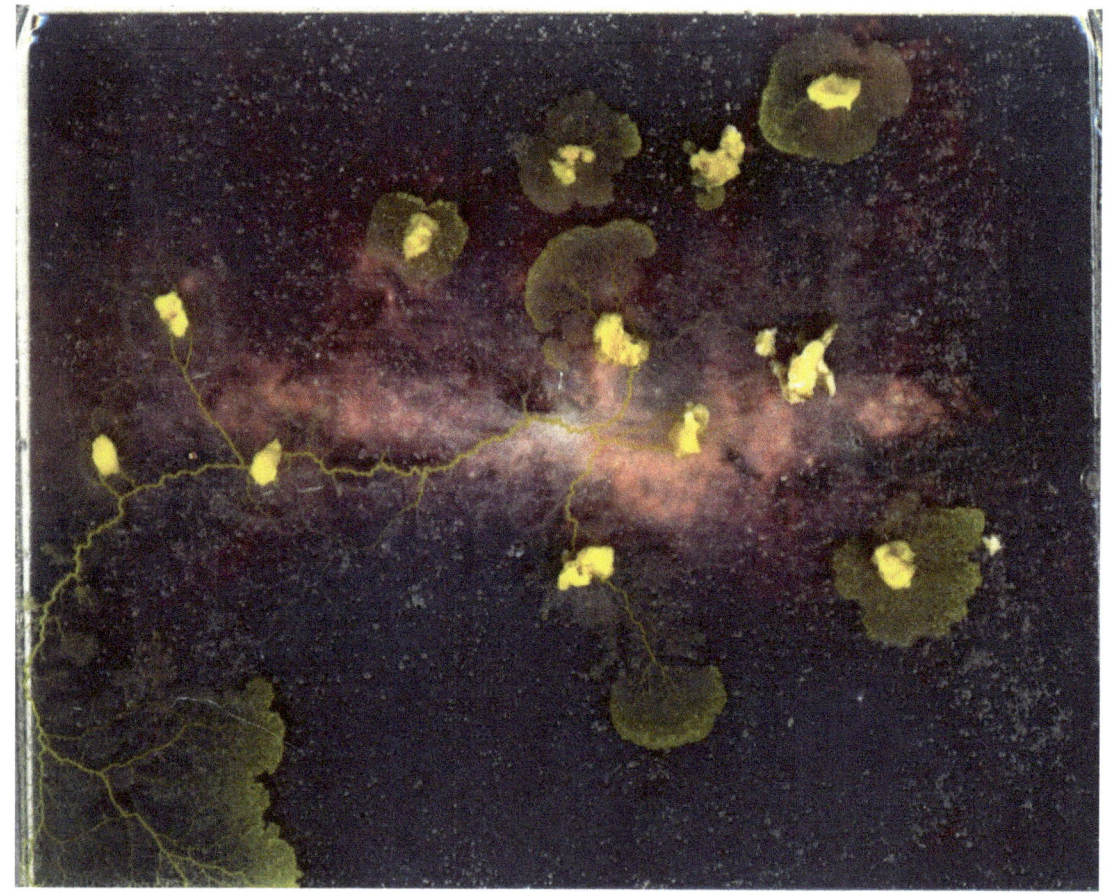

Milky Way.
60×50 cm. Direct printing on brush-finished aluminium.

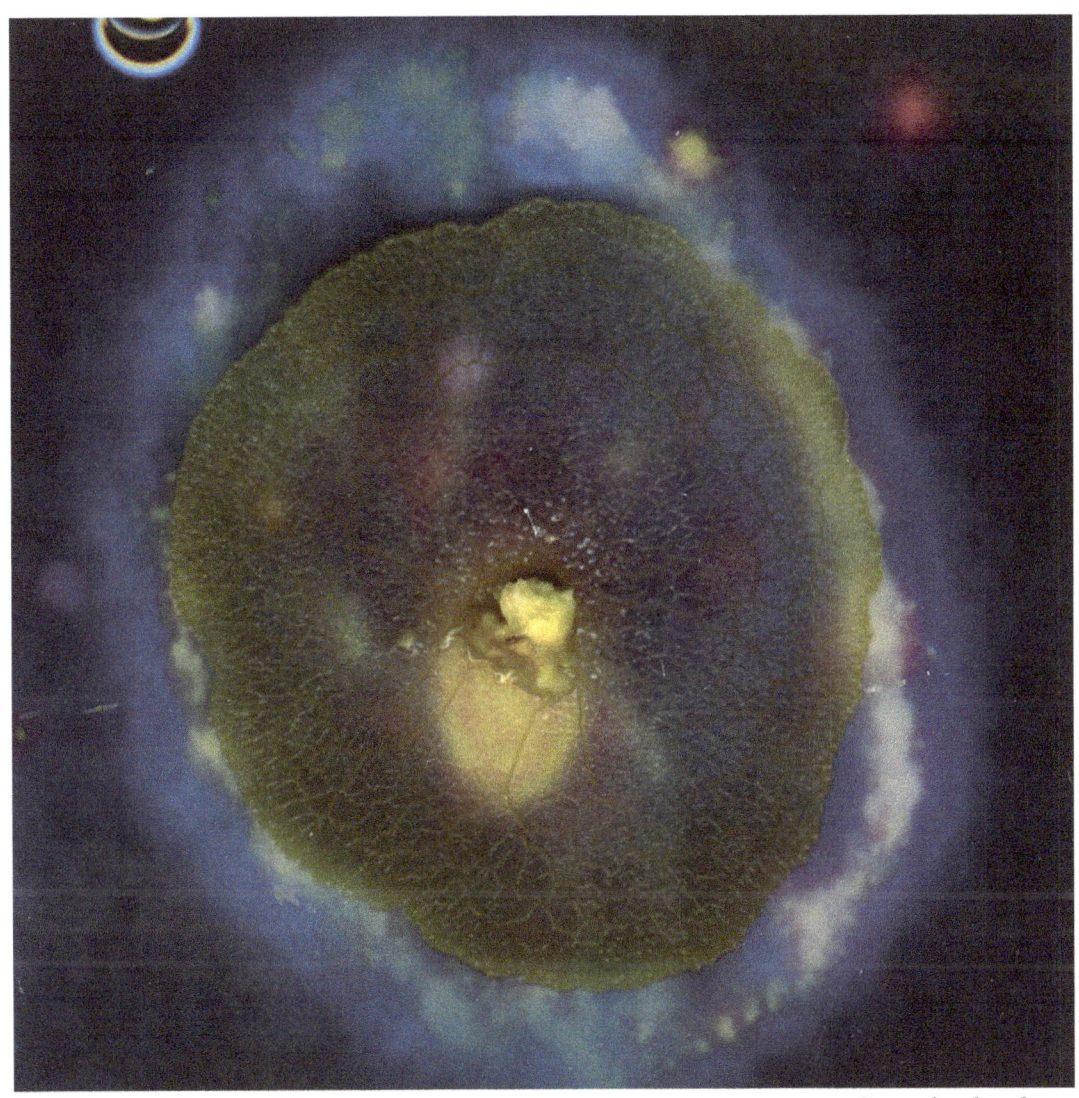

Cartwheel galaxy.
40×40 cm. Direct printing on brush-finished aluminium.

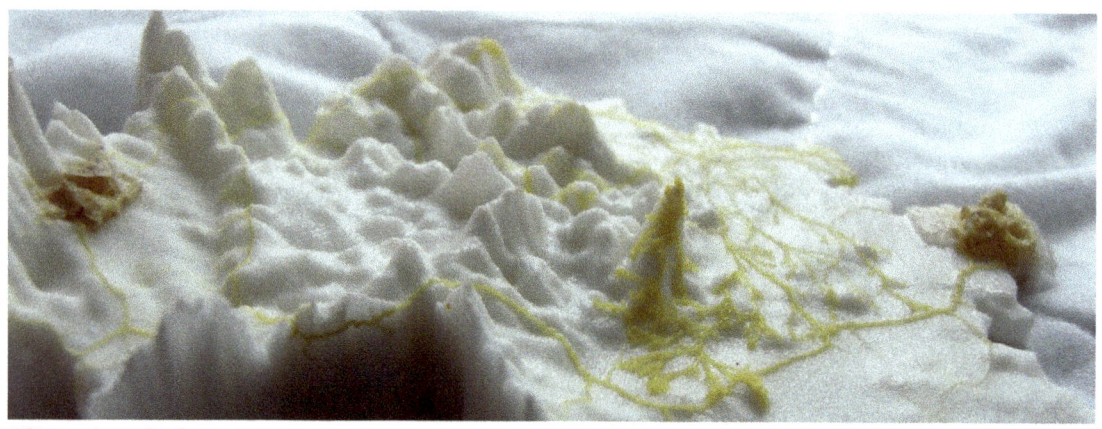

Slimy Autobahns.
90×30 cm. Direct printing on brush-finished aluminium.

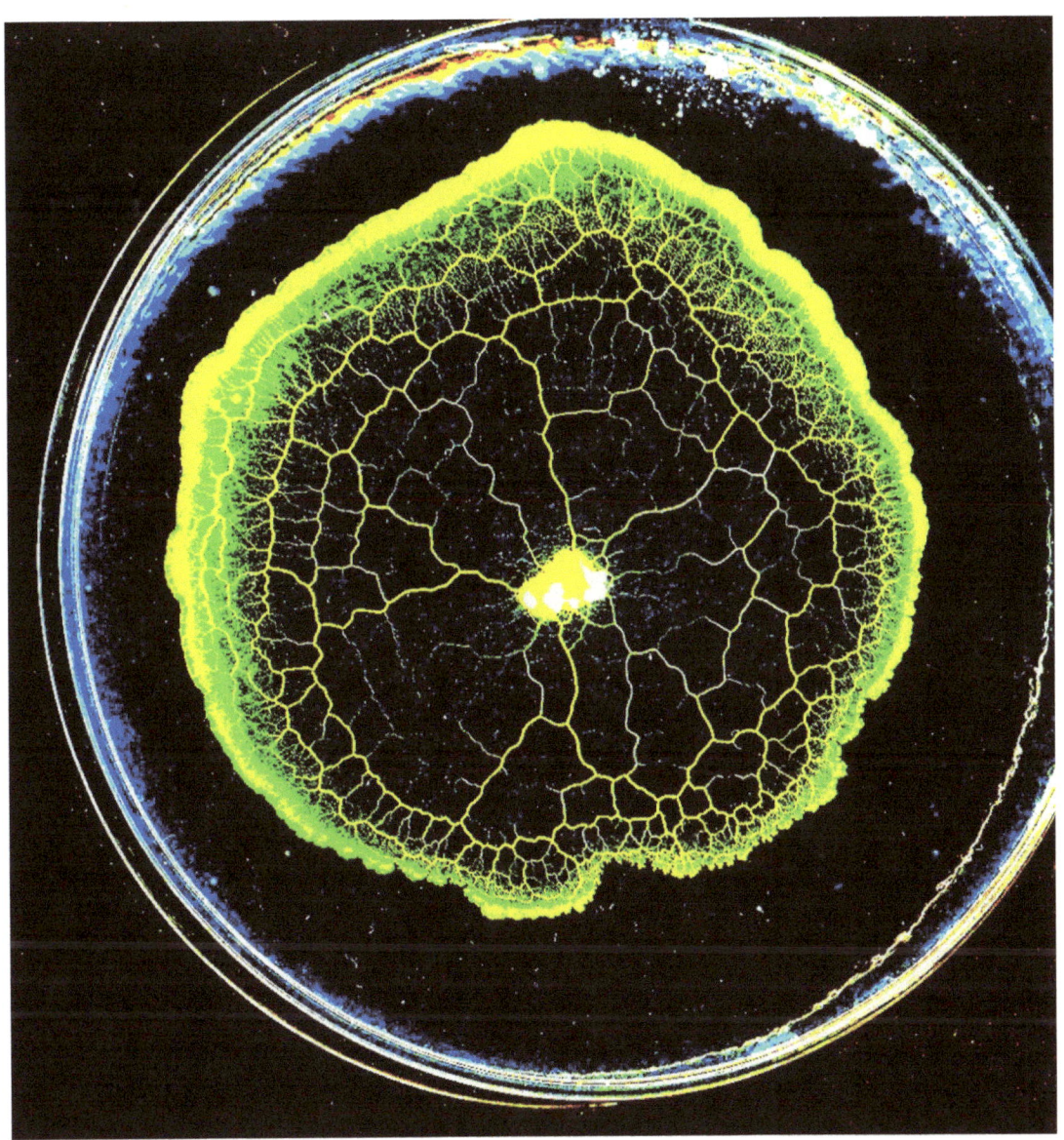

Nutrient rich substrate.
40×40 cm. Direct printing on brush-finished aluminium.

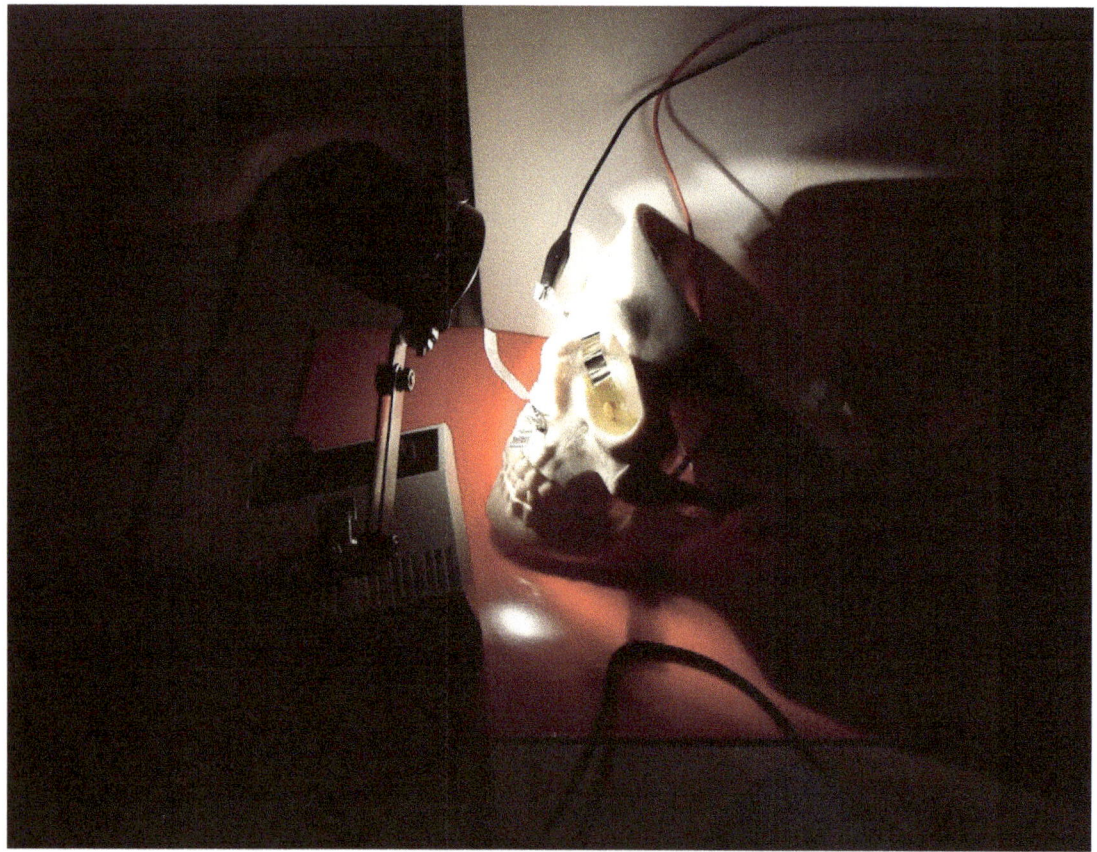

Slime eye.
60×45 cm. Direct printing on brush-finished aluminium.

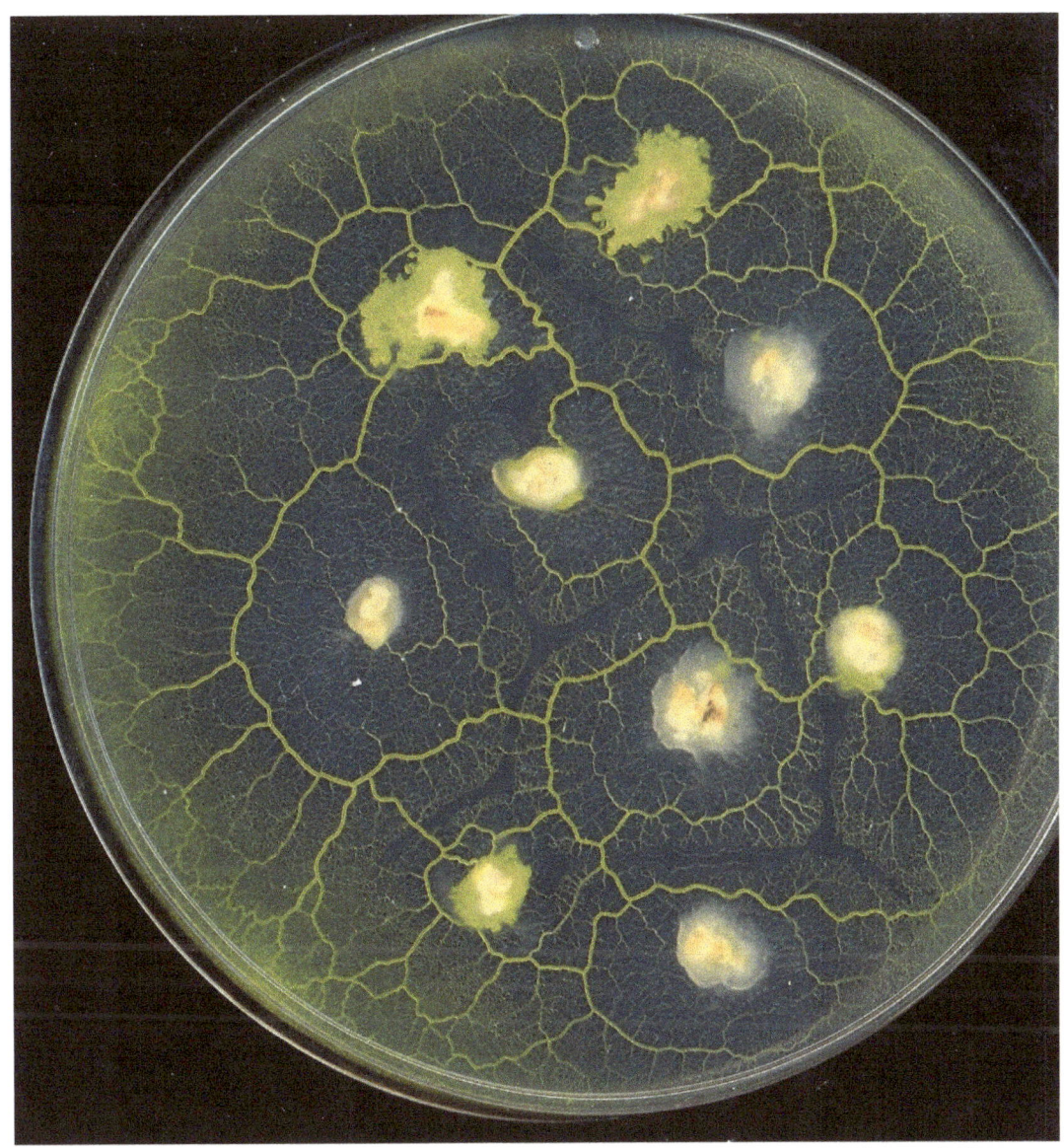

Voronoi diagram.
40×40 cm. Direct printing on brush-finished aluminium.

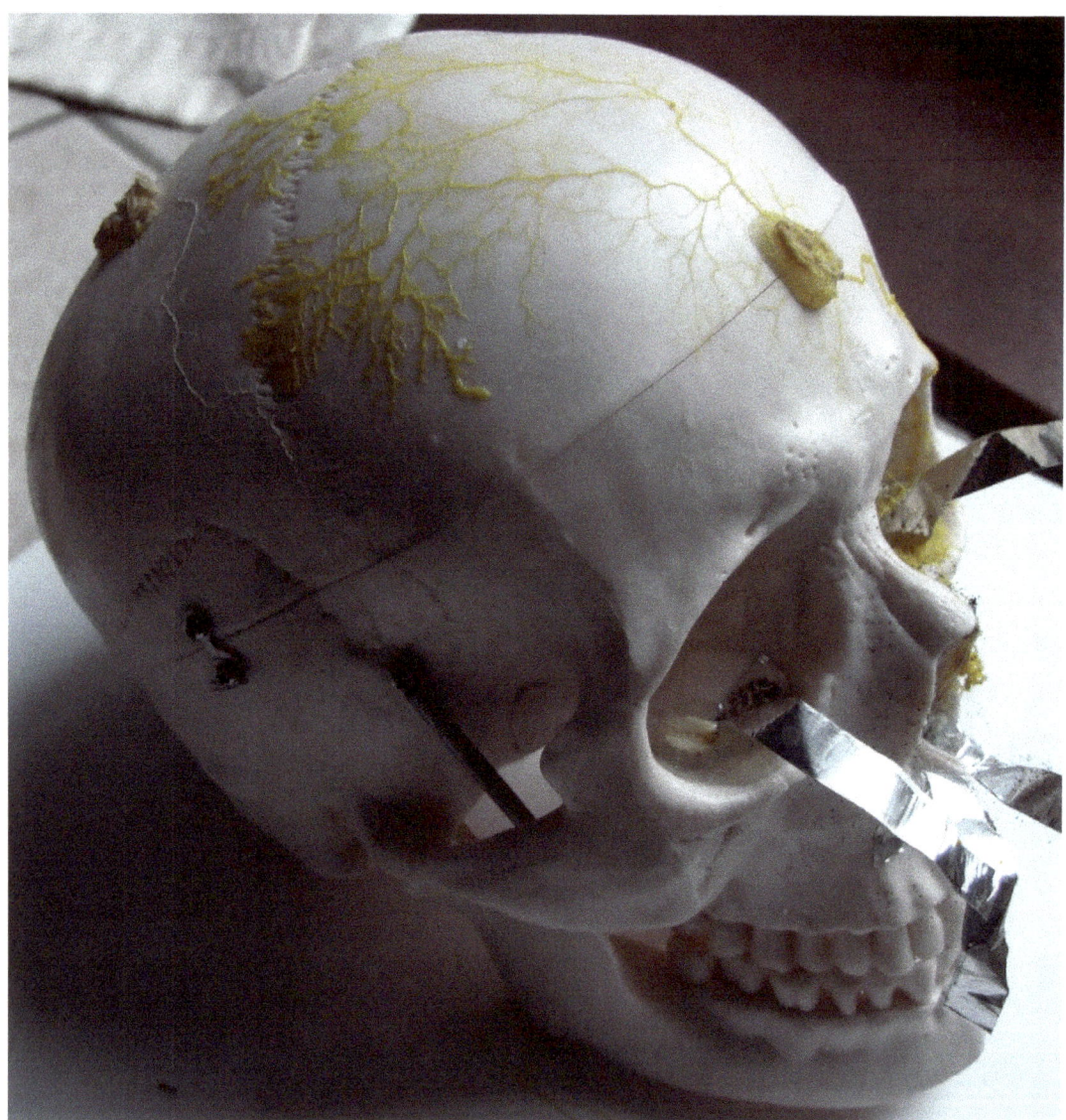

Innervation I.
40×40 cm. Direct printing on brush-finished aluminium.

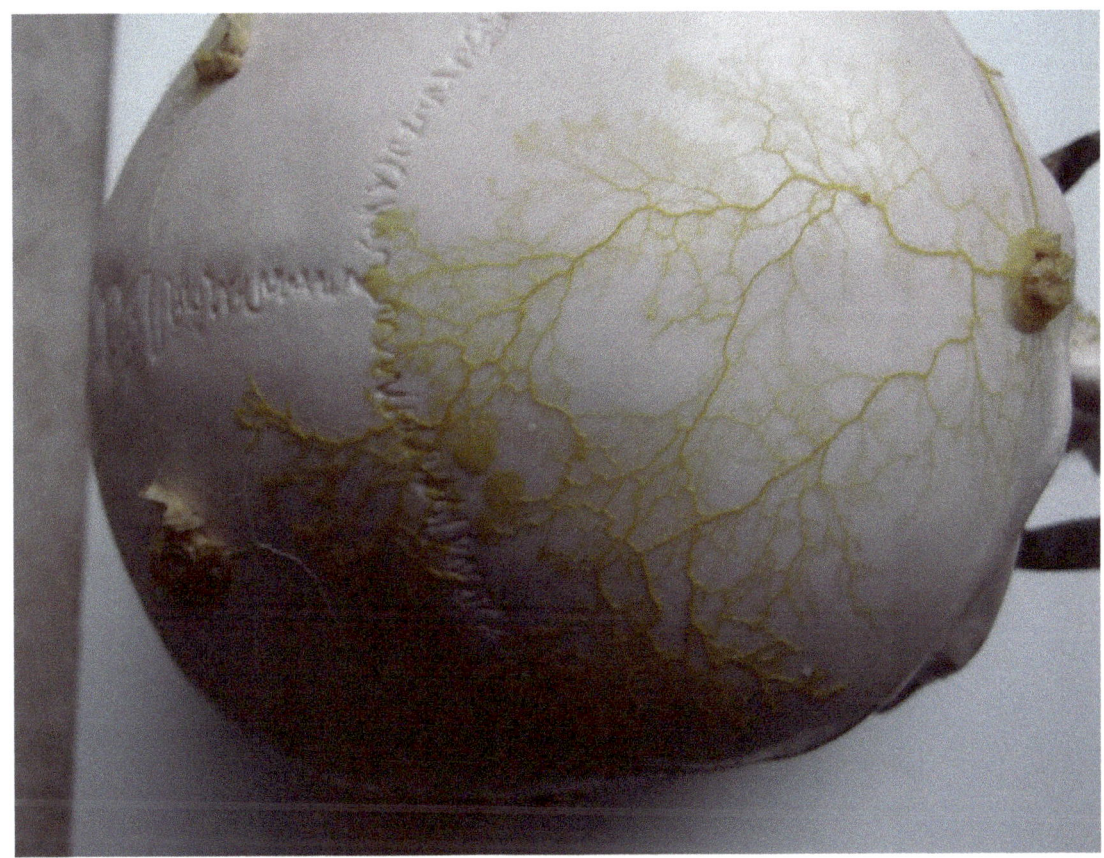

Innervation II.
60×45 cm. Direct printing on brush-finished aluminium.

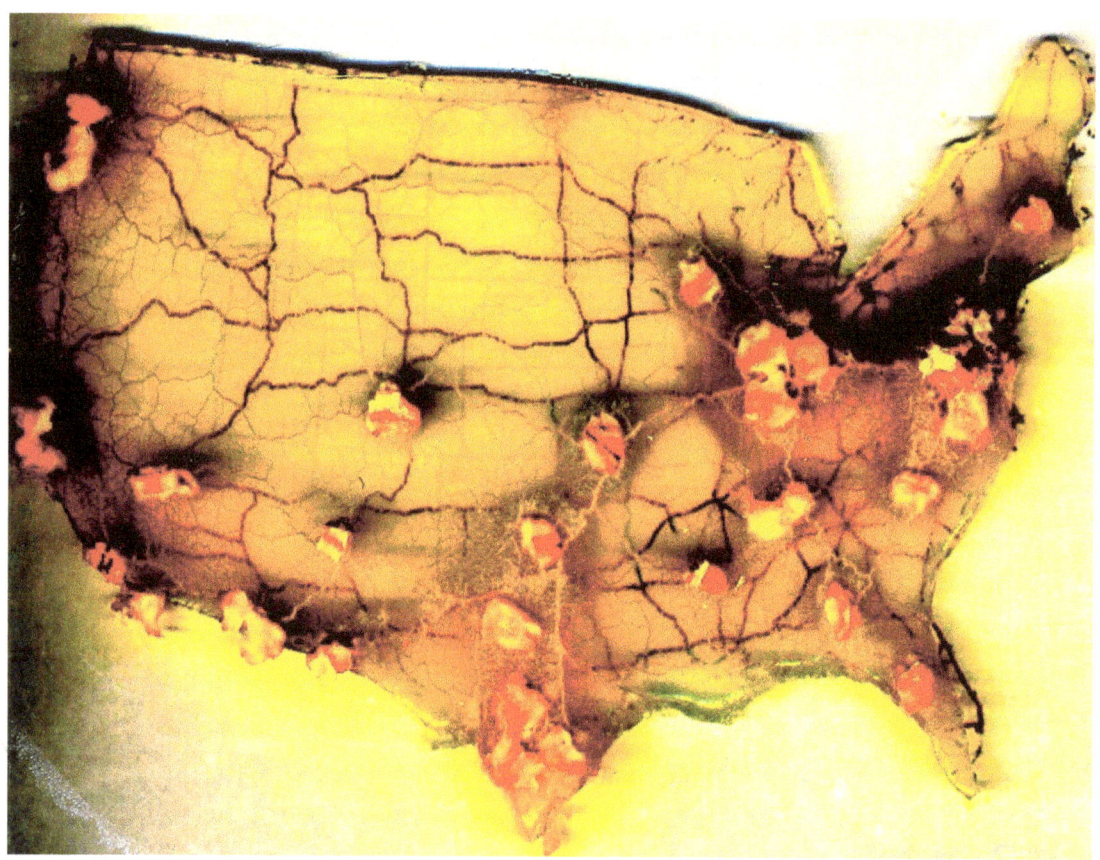

Highways II.
64×36 cm. Direct printing on brush-finished aluminium.

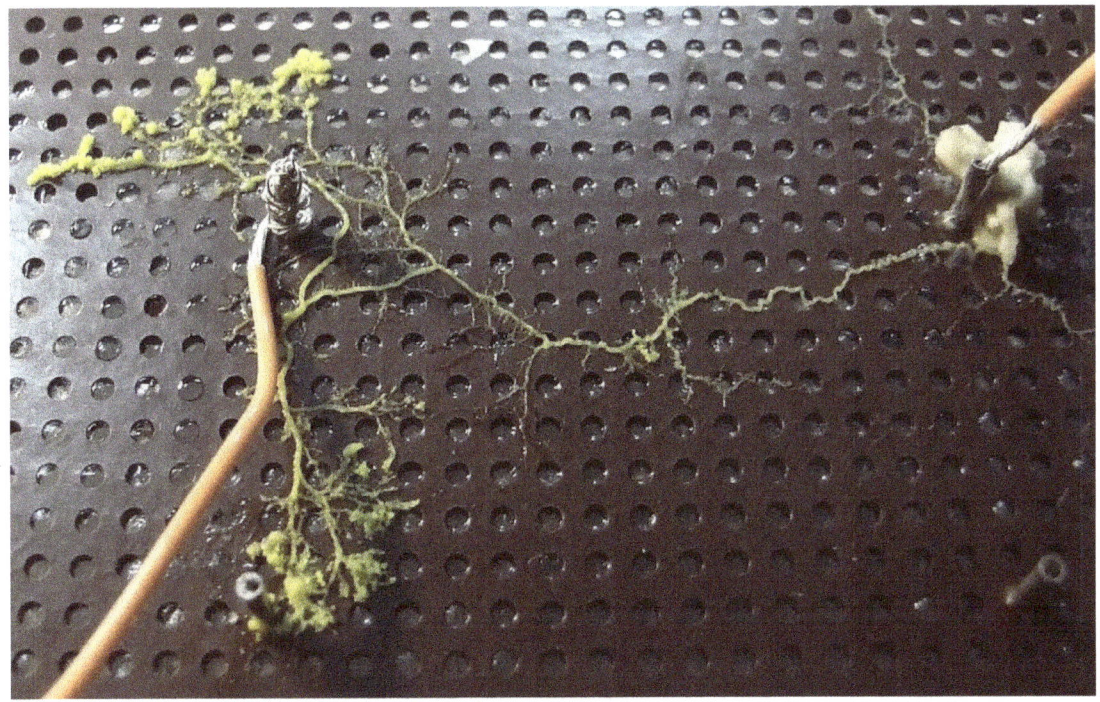

Slime wires III.
64×36 cm. Direct printing on brush-finished aluminium.

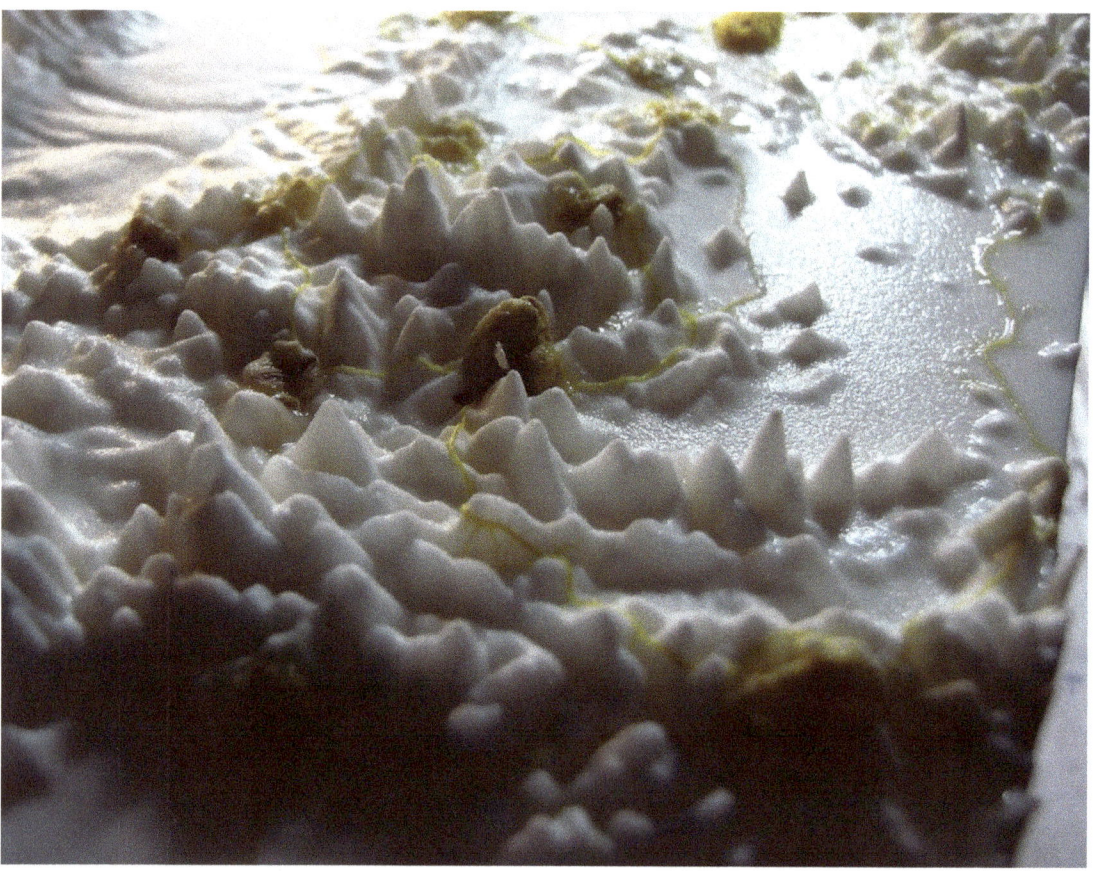

Balkans 3D.
40×30 cm. Direct printing on brush-finished aluminium.

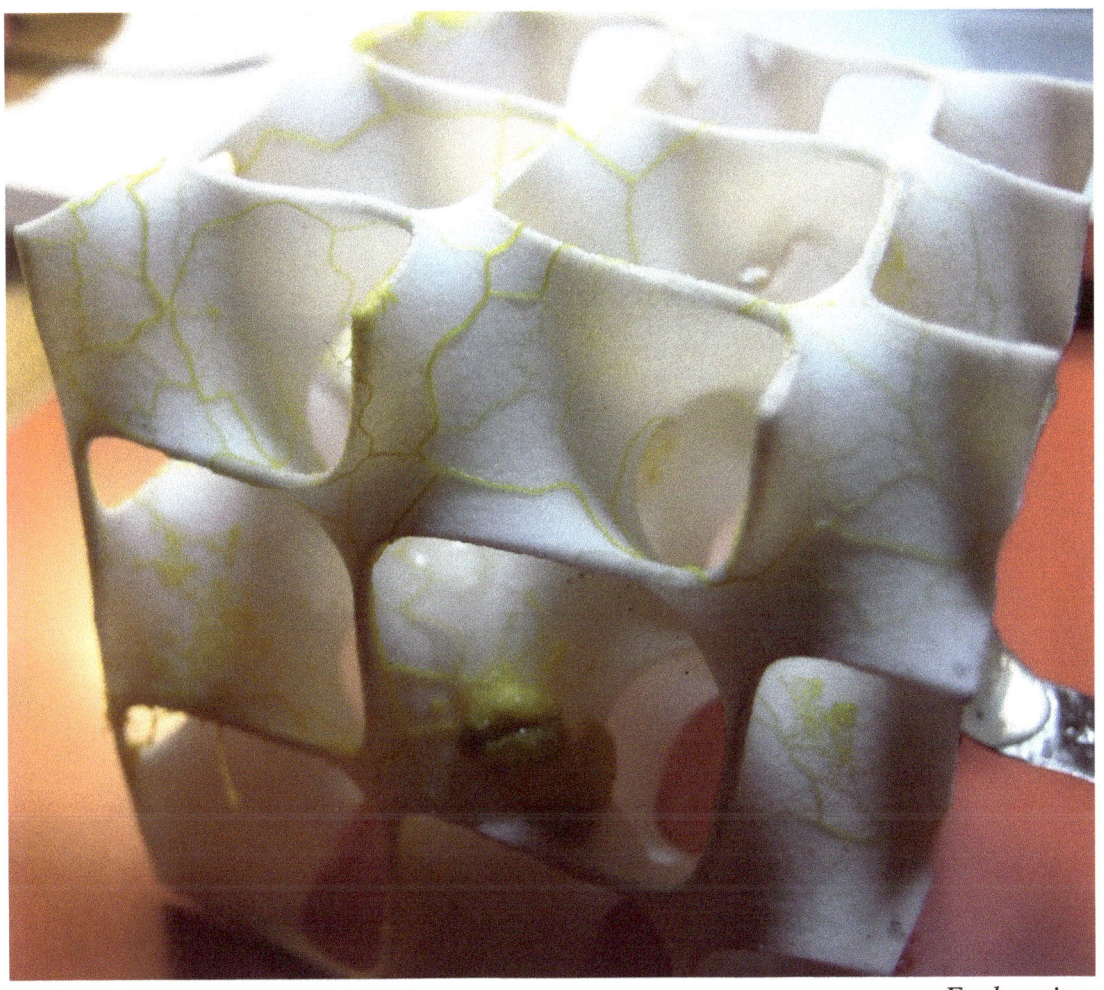

Exploration.
30×24 cm. Direct printing on brush-finished aluminium.

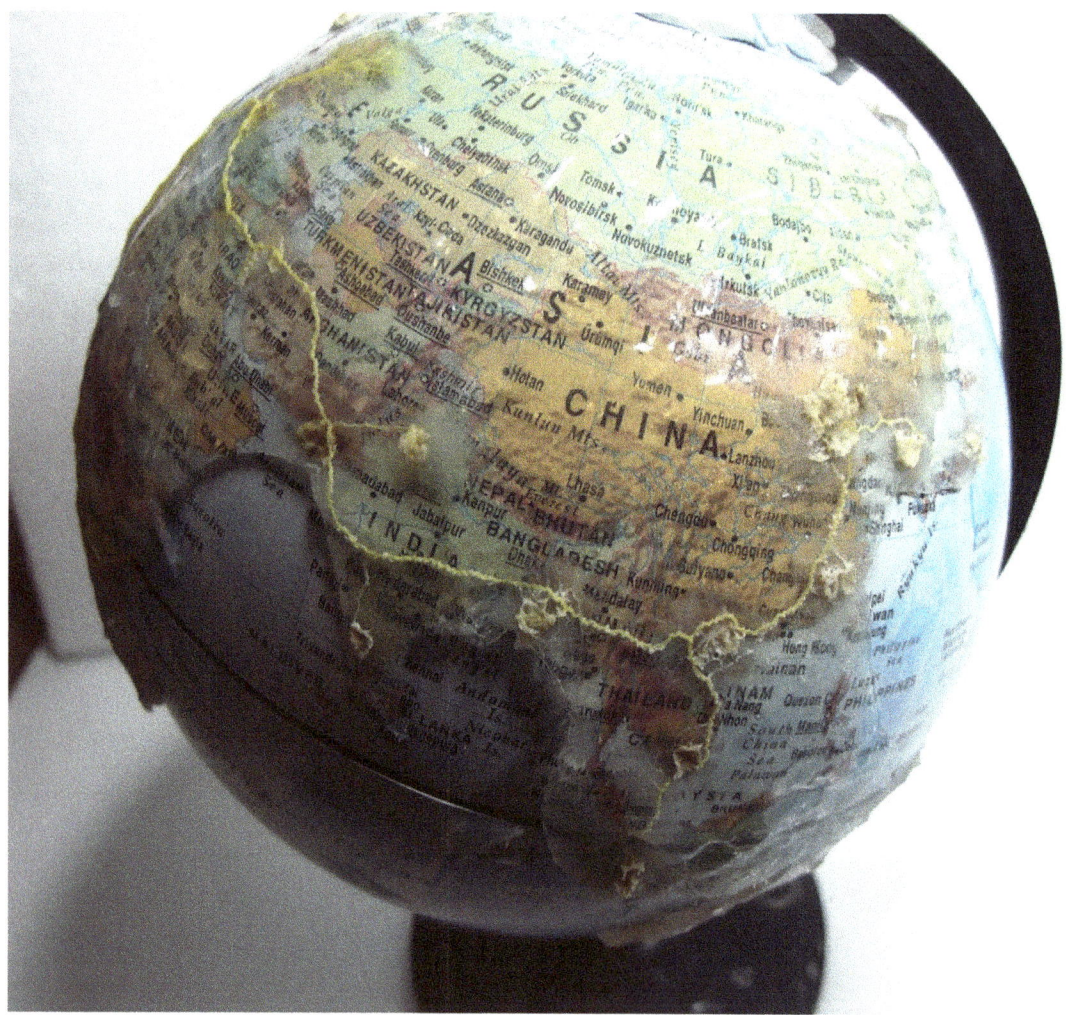

World colonisation I. The Silk Road.
40×40 cm. Direct printing on brush-finished aluminium.

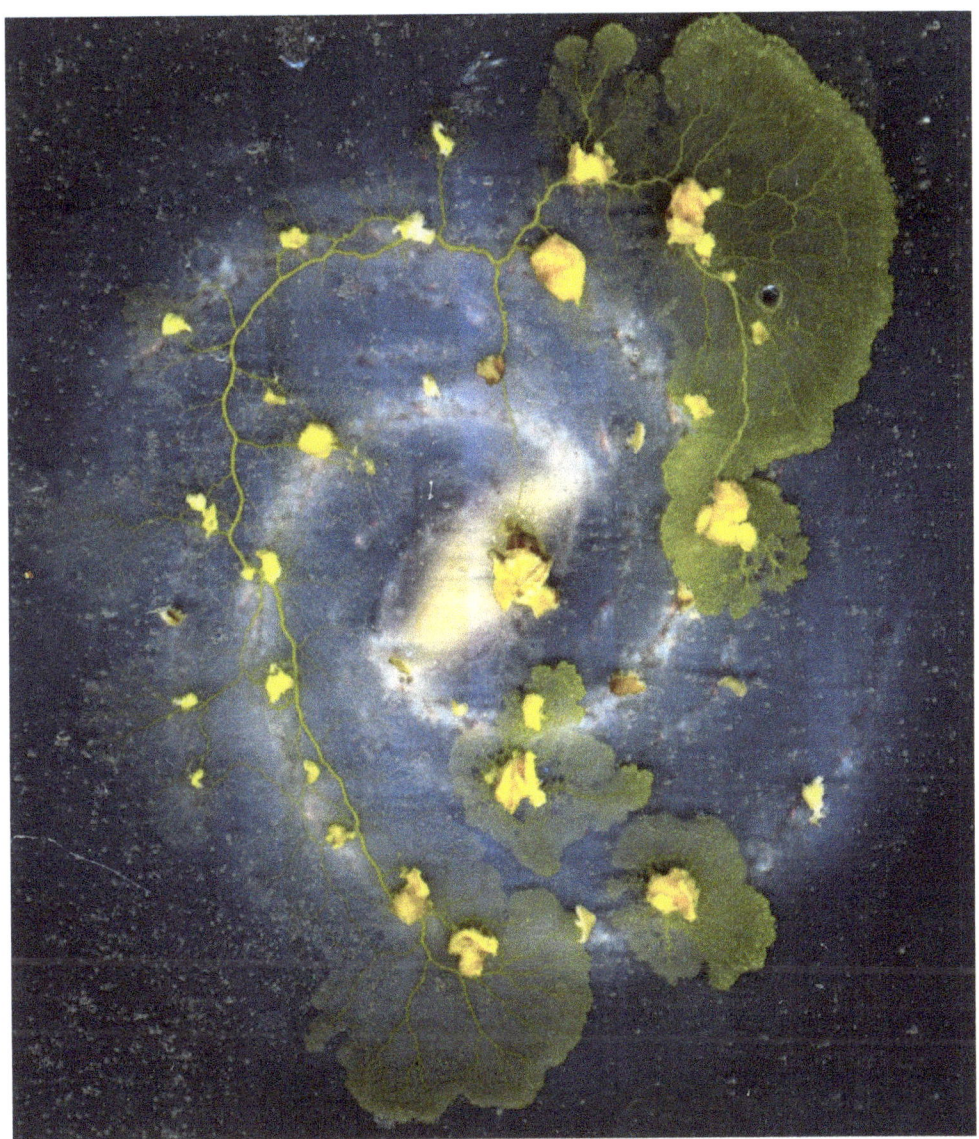

Galaxy II.
50×60 cm. Direct printing on brush-finished aluminium.

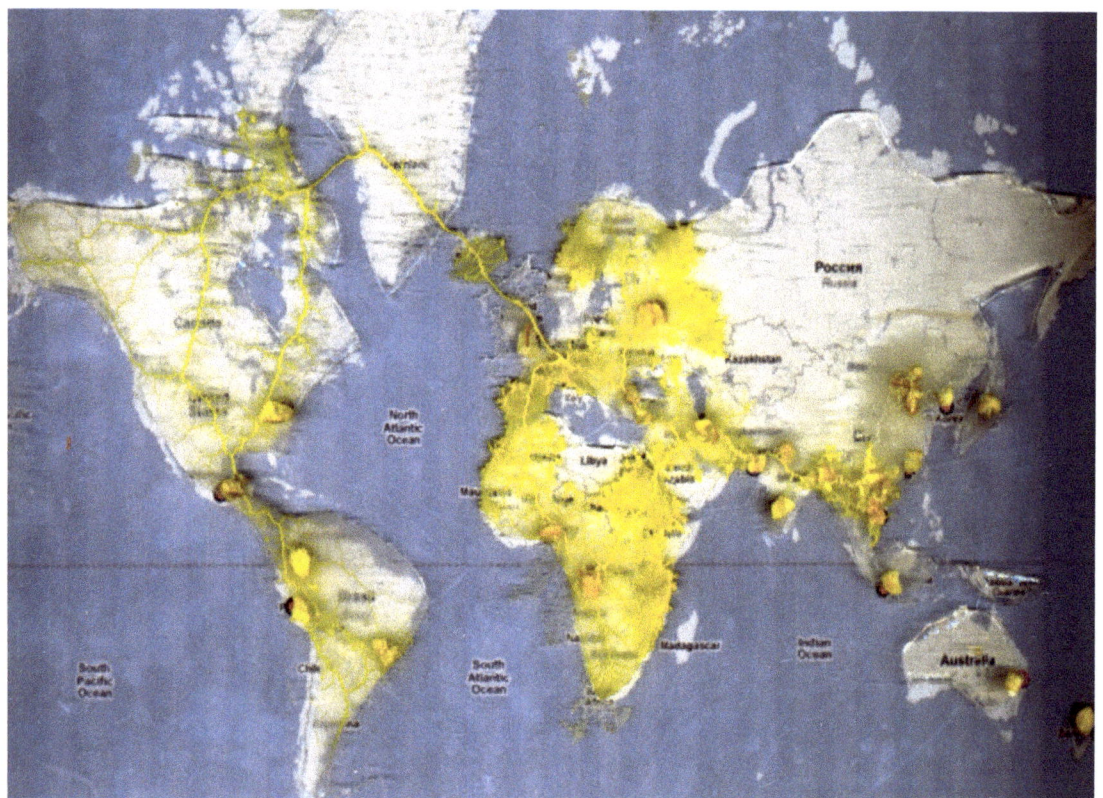

World colonisation II.
40×30 cm. Direct printing on brush-finished aluminium.

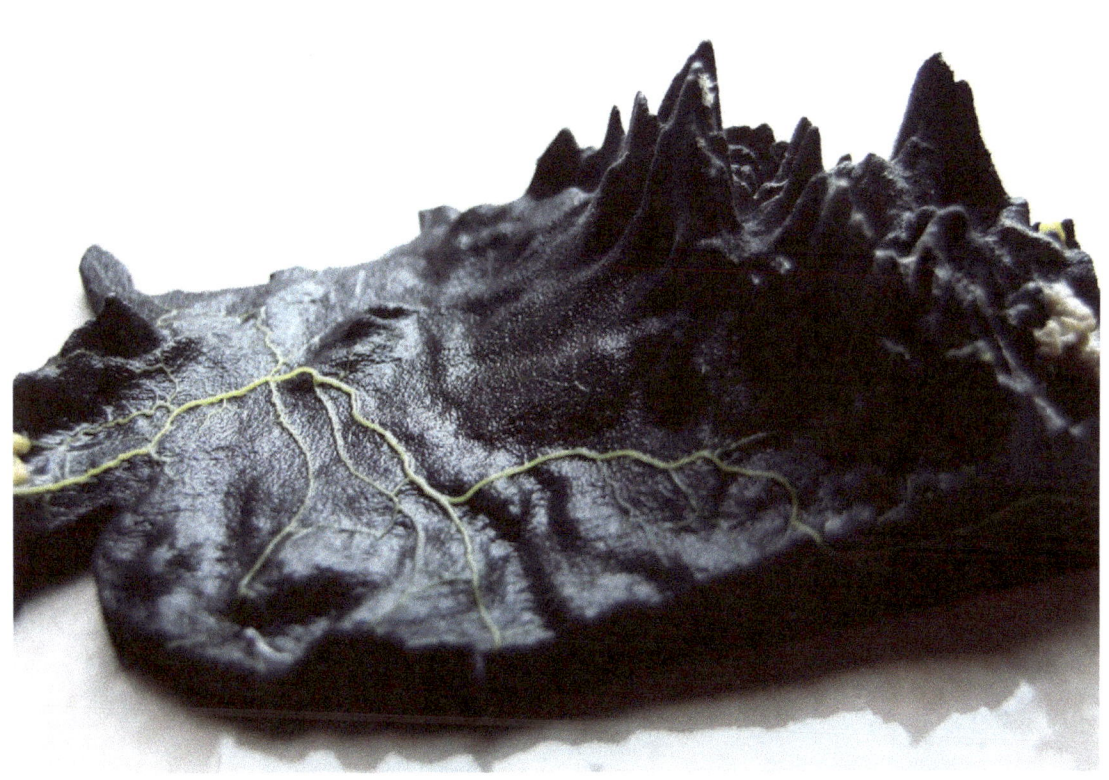

Route planning.
40×30 cm. Direct printing on brush-finished aluminium.

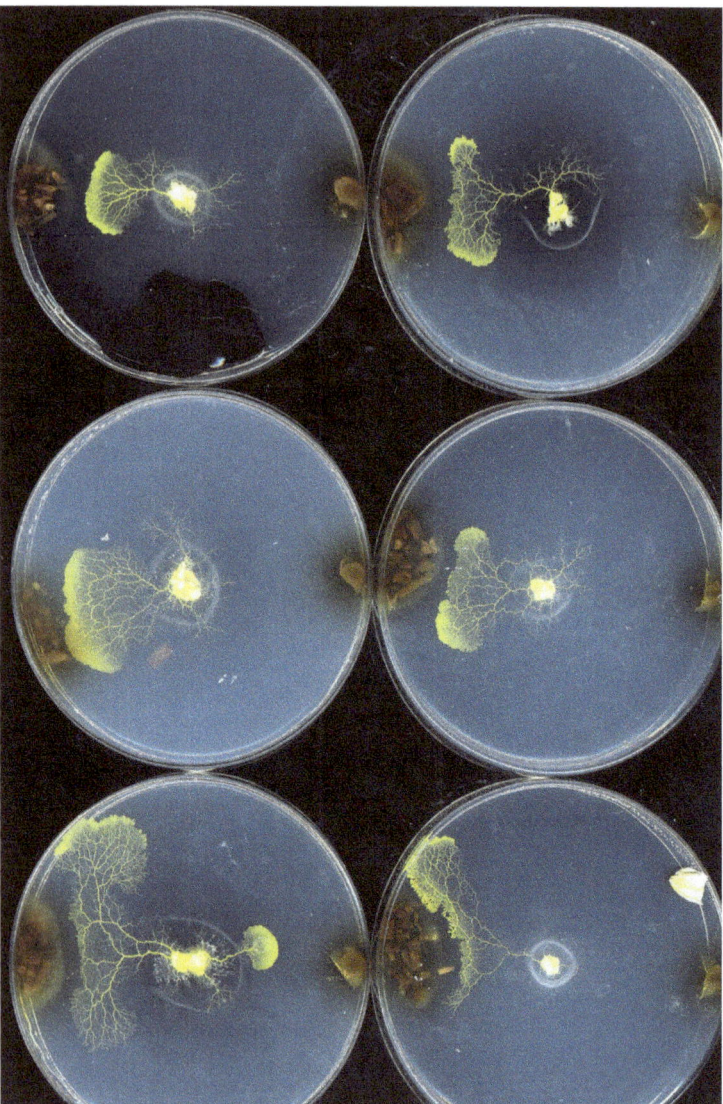

Valerian versus Kalms Sleep.
45×60 cm. Direct printing on brush-finished aluminium.

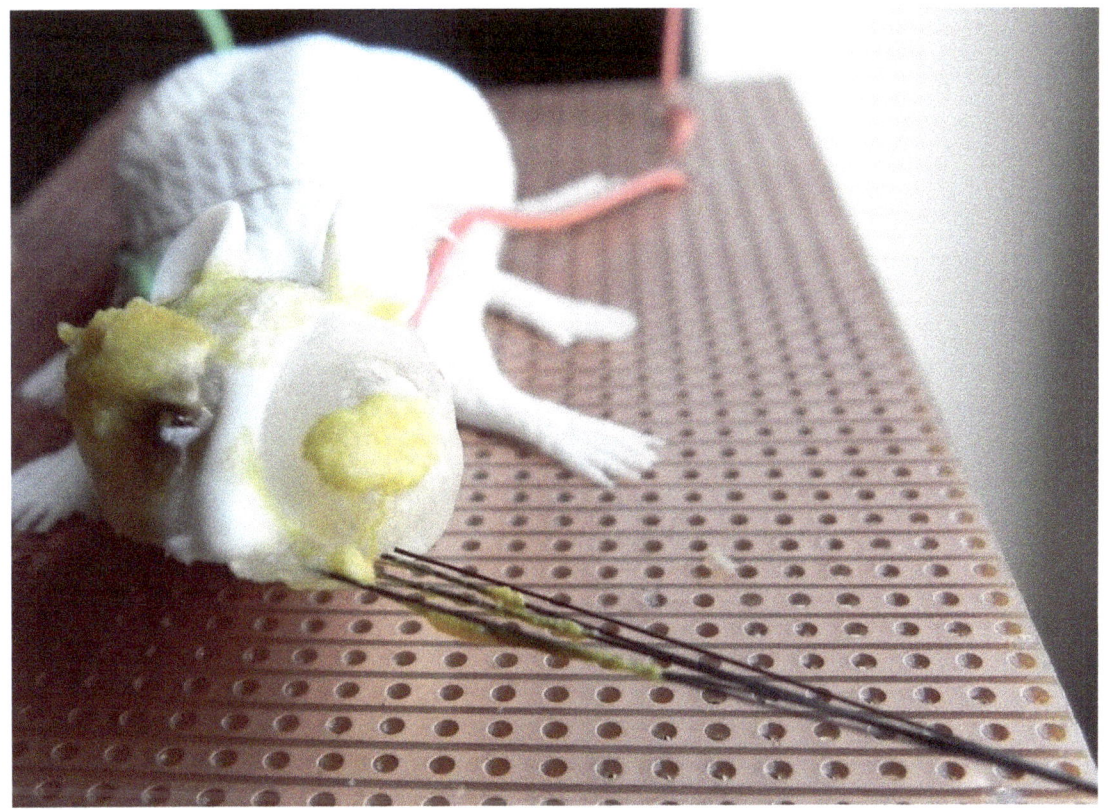

Slime whiskers.
60×45 cm. Direct printing on brush-finished aluminium.

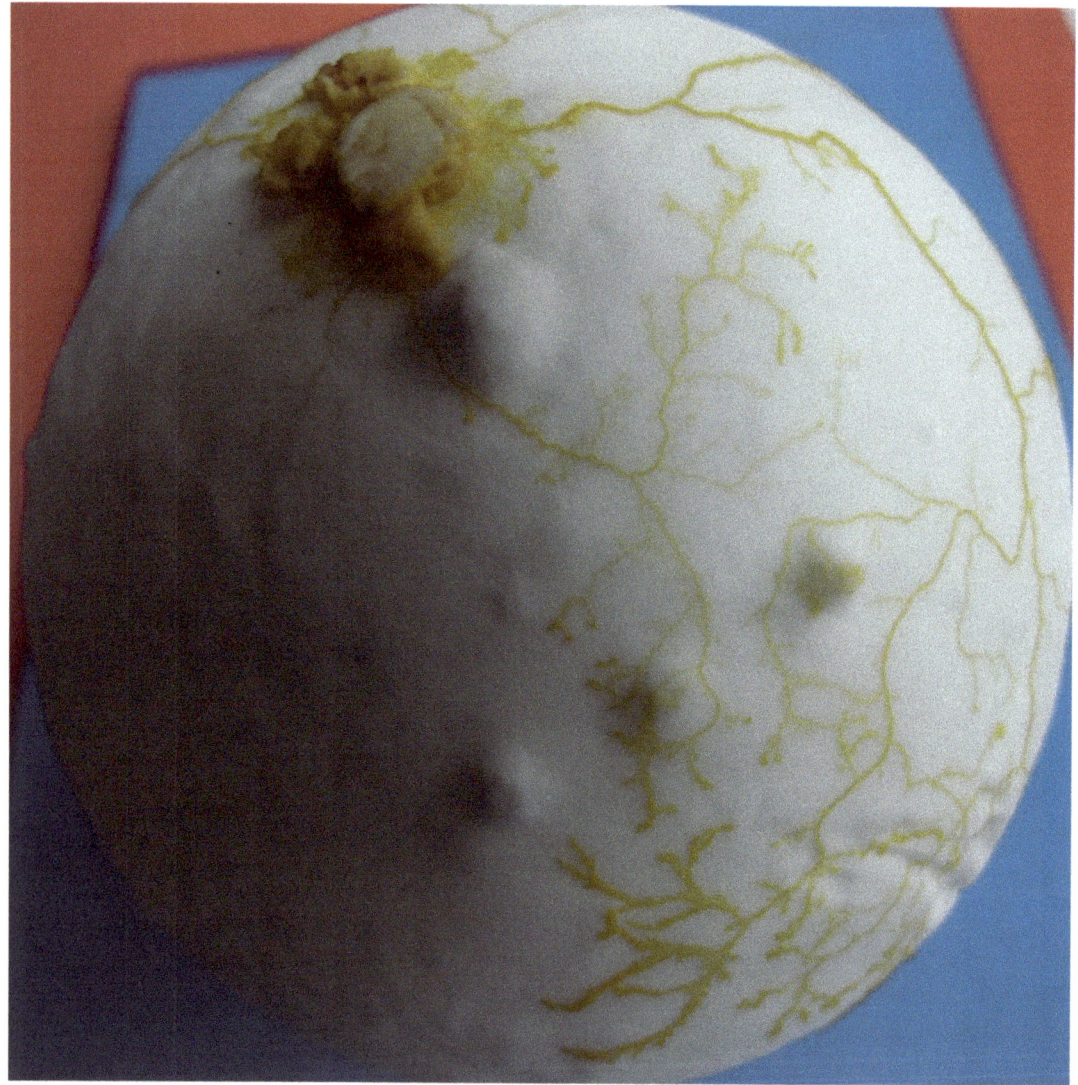

Colonisation of Mars.
40×40 cm. Direct printing on brush-finished aluminium.

The Silence of Slime Mould

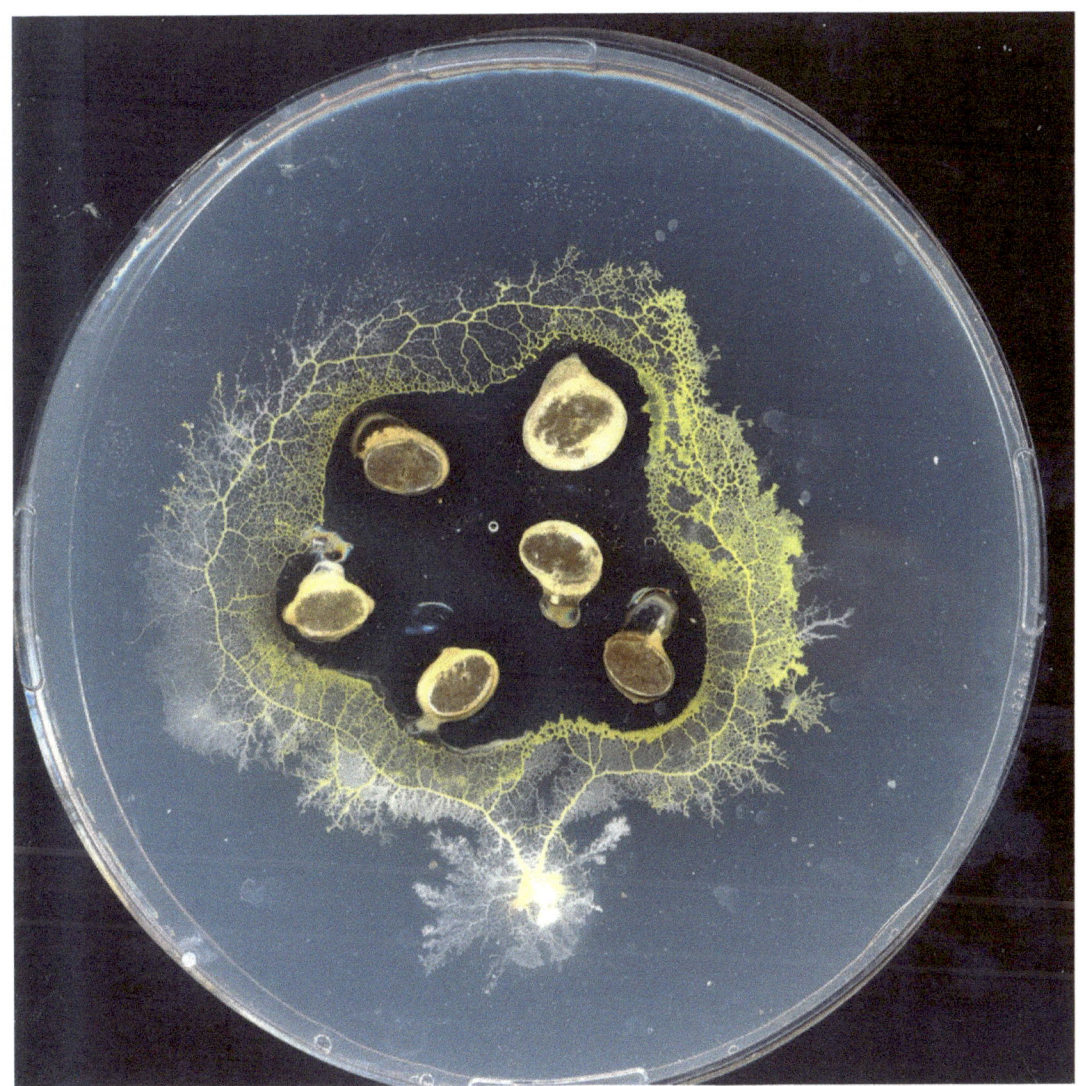

Concave hull.
40×40 cm. Direct printing on brush-finished aluminium.

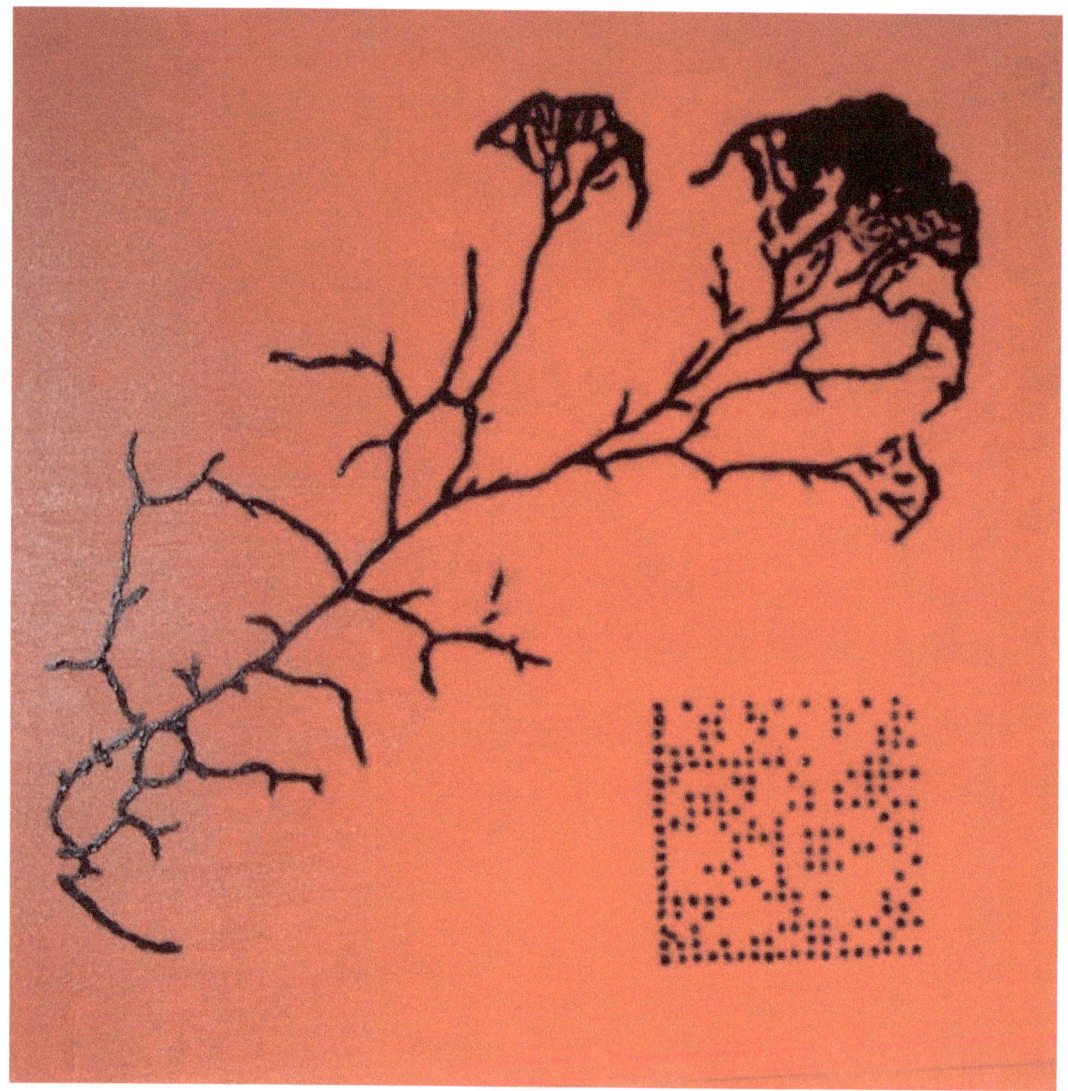

Active zone.
50×50 cm. Acrylic paints. Canvas.

The Silence of Slime Mould

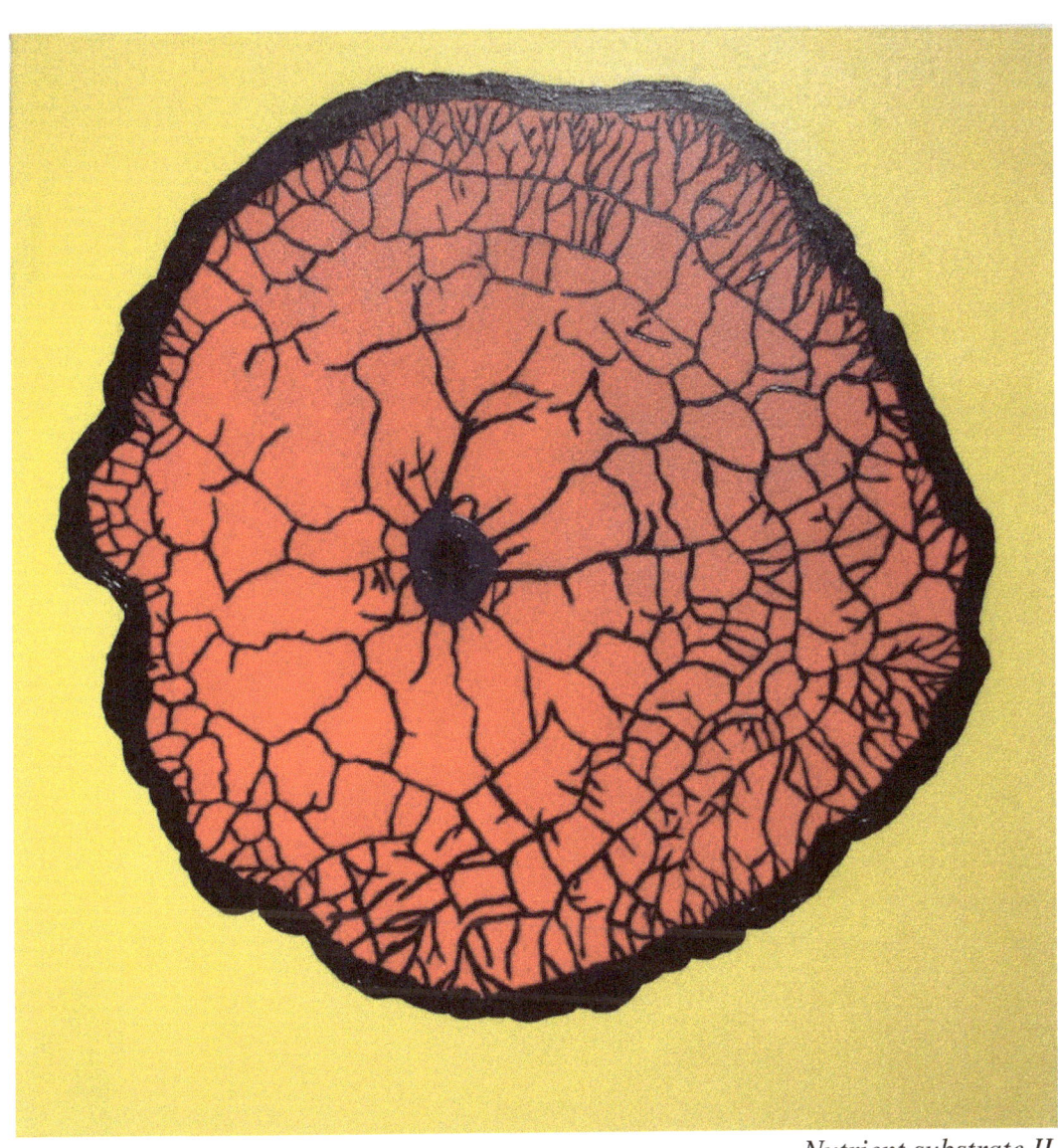

Nutrient substrate II.
50×50 cm. Acrylic paints. Canvas.

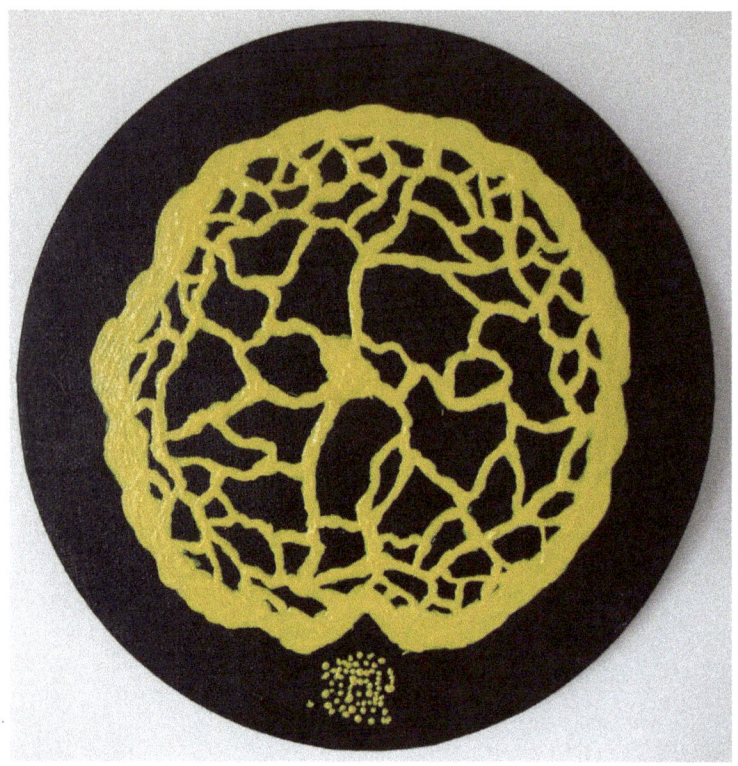

Propagation.
Round canvas 20 cm diameter. Acrylic paints. Canvas.

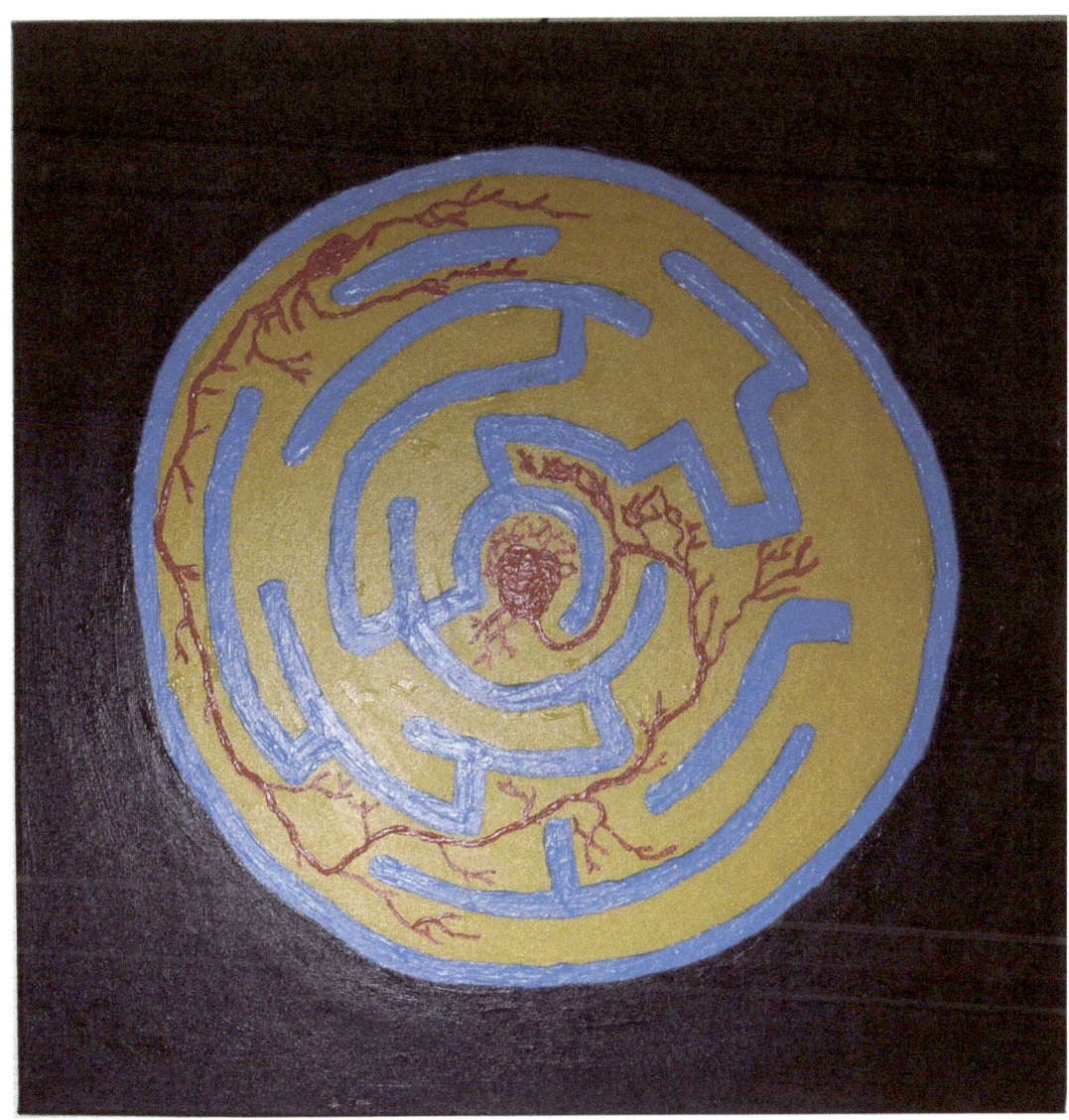

Maze II.
50×50 cm. Acrylic paints. Canvas.

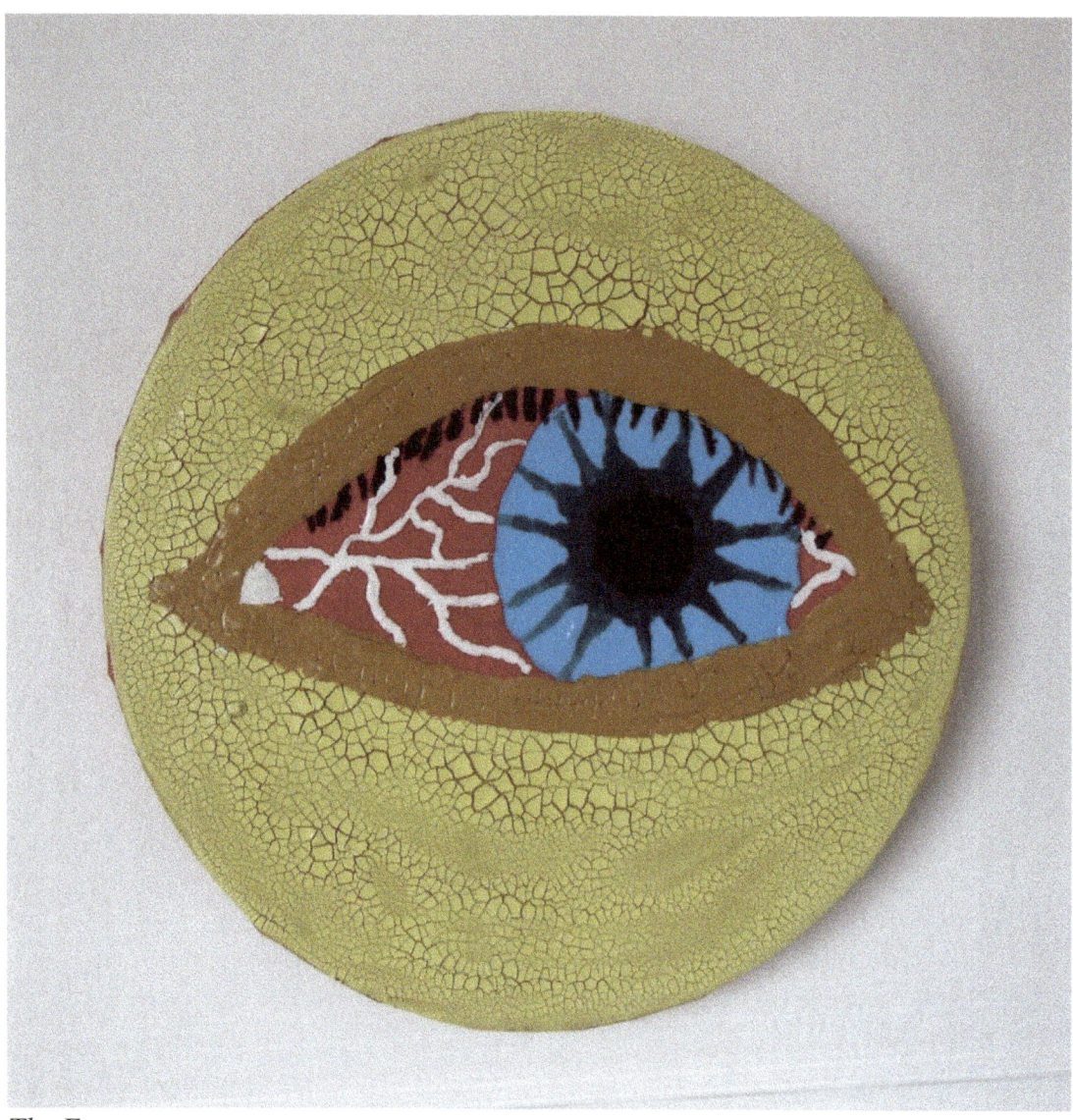

The Eye.
20 cm diameter. Acrylic paints. Round canvas.

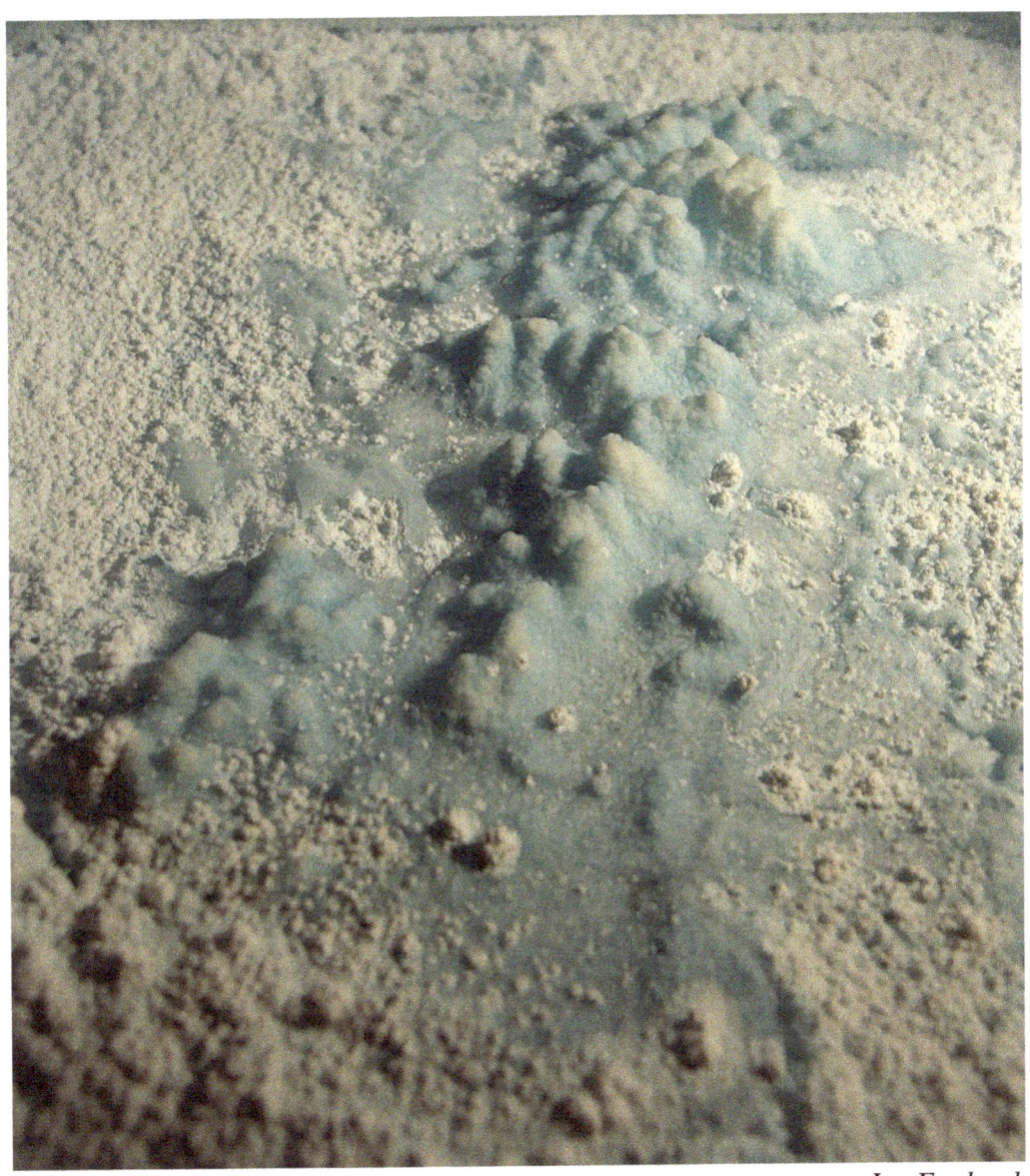

Ice England.
40×40 cm. Prints under acrylic glass.

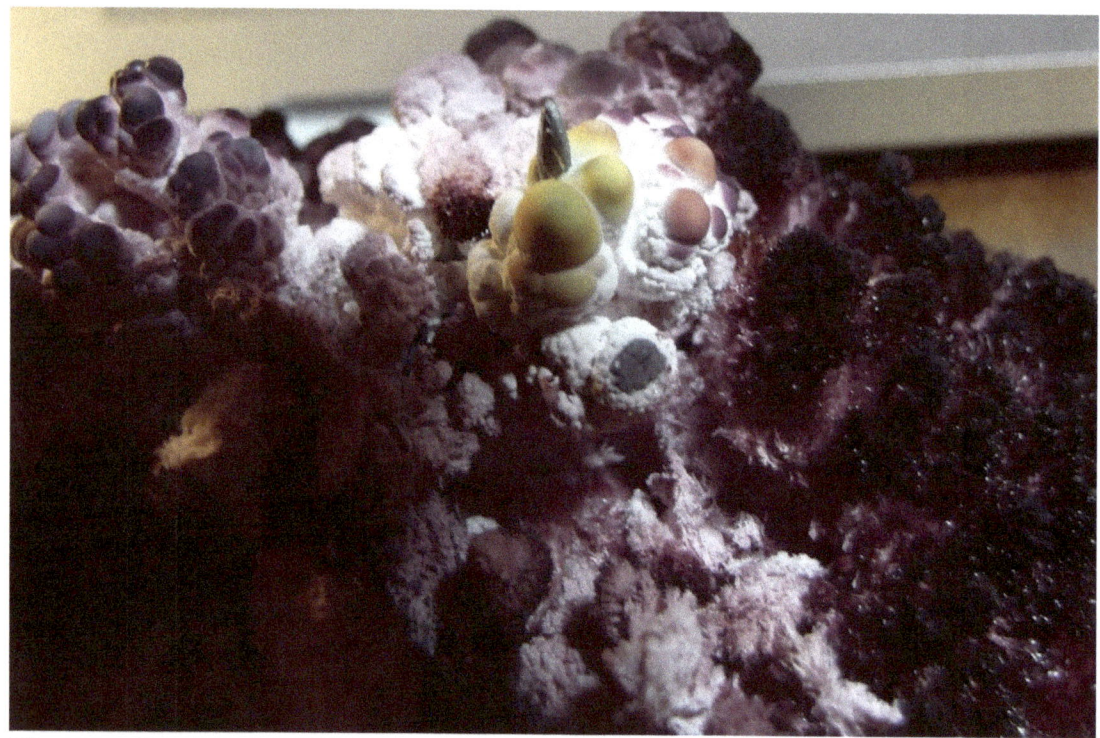

The growth.
75×50 cm. Prints under acrylic glass.

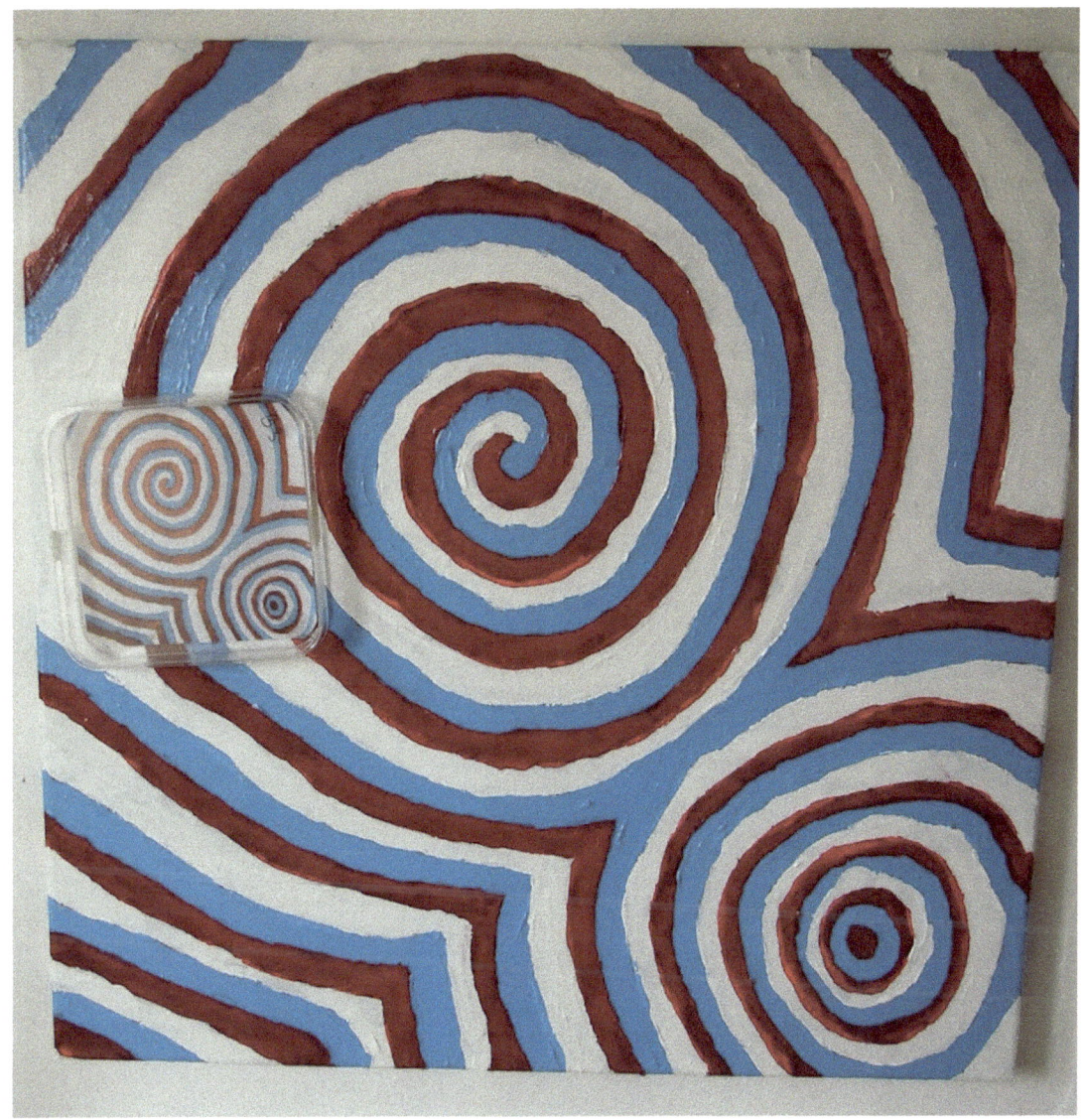

Excitation waves.
50×50 cm. Canvas. Acrylic paints. Square Petri dish 12×12 cm is installed on the canvas.

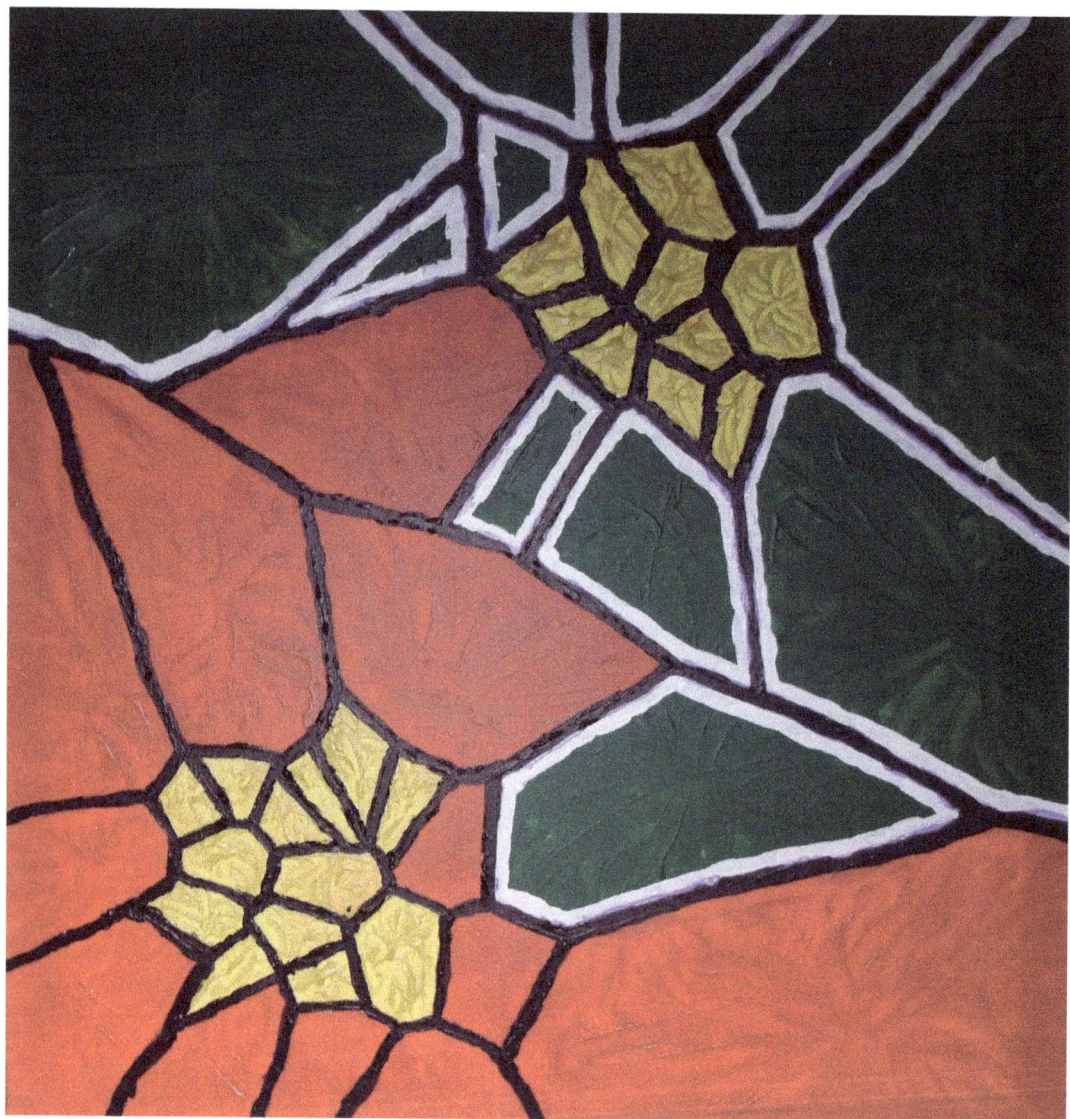

Voronoi diagram II.
60×60 cm. Canvas. Acrylic paints.

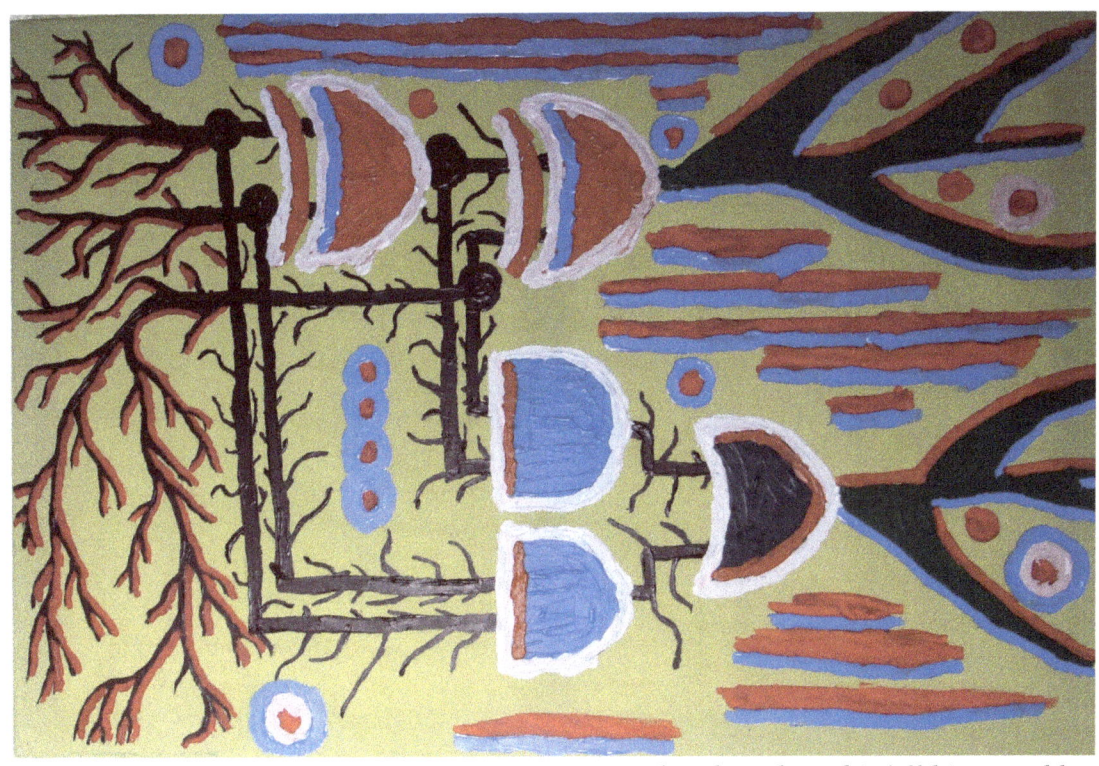

Plant based one bit full binary adder.
50×60 cm. Canvas. Acrylic paints.

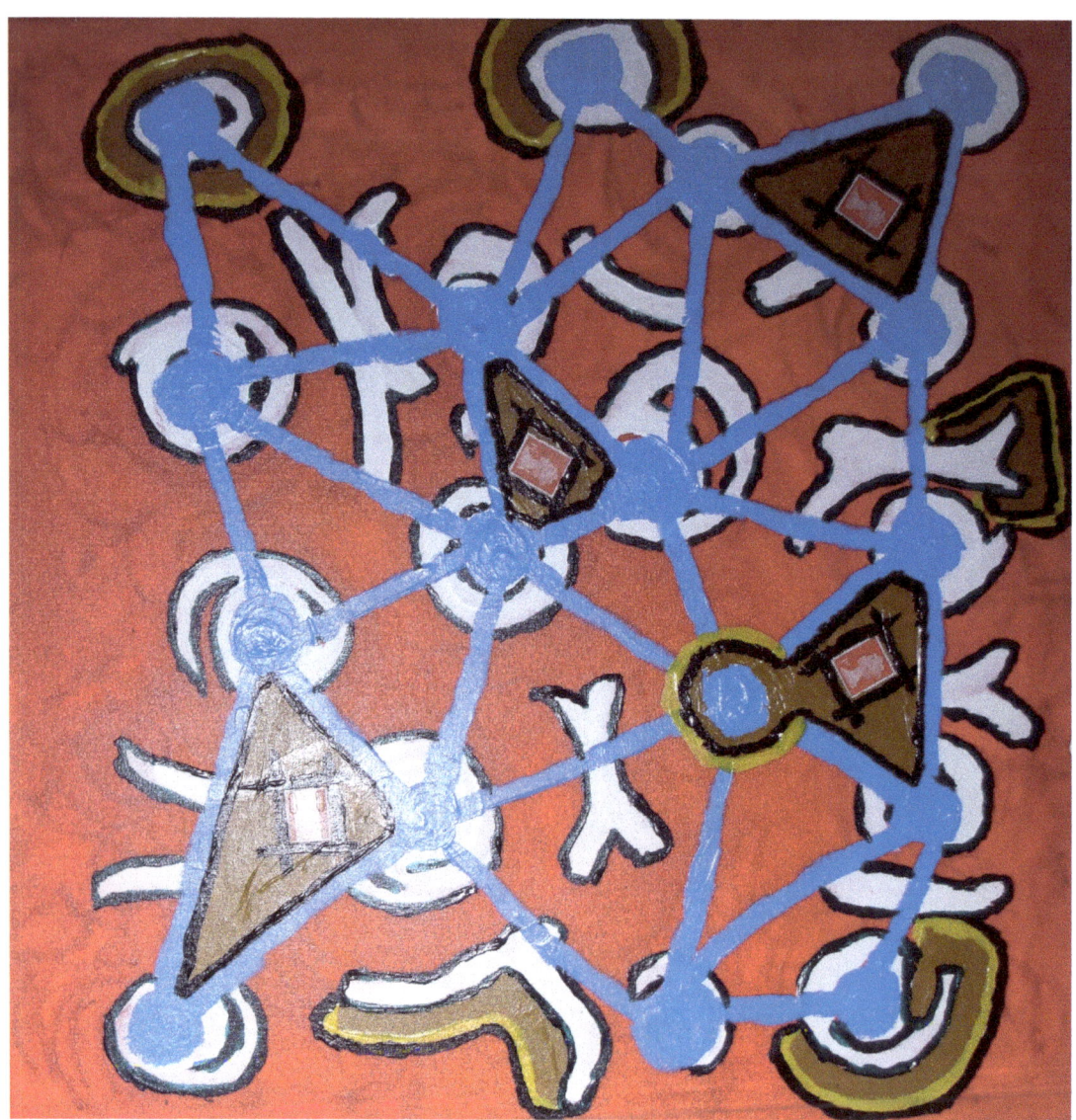

Proximity graph.
60×60 cm. Canvas. Acrylic paints.

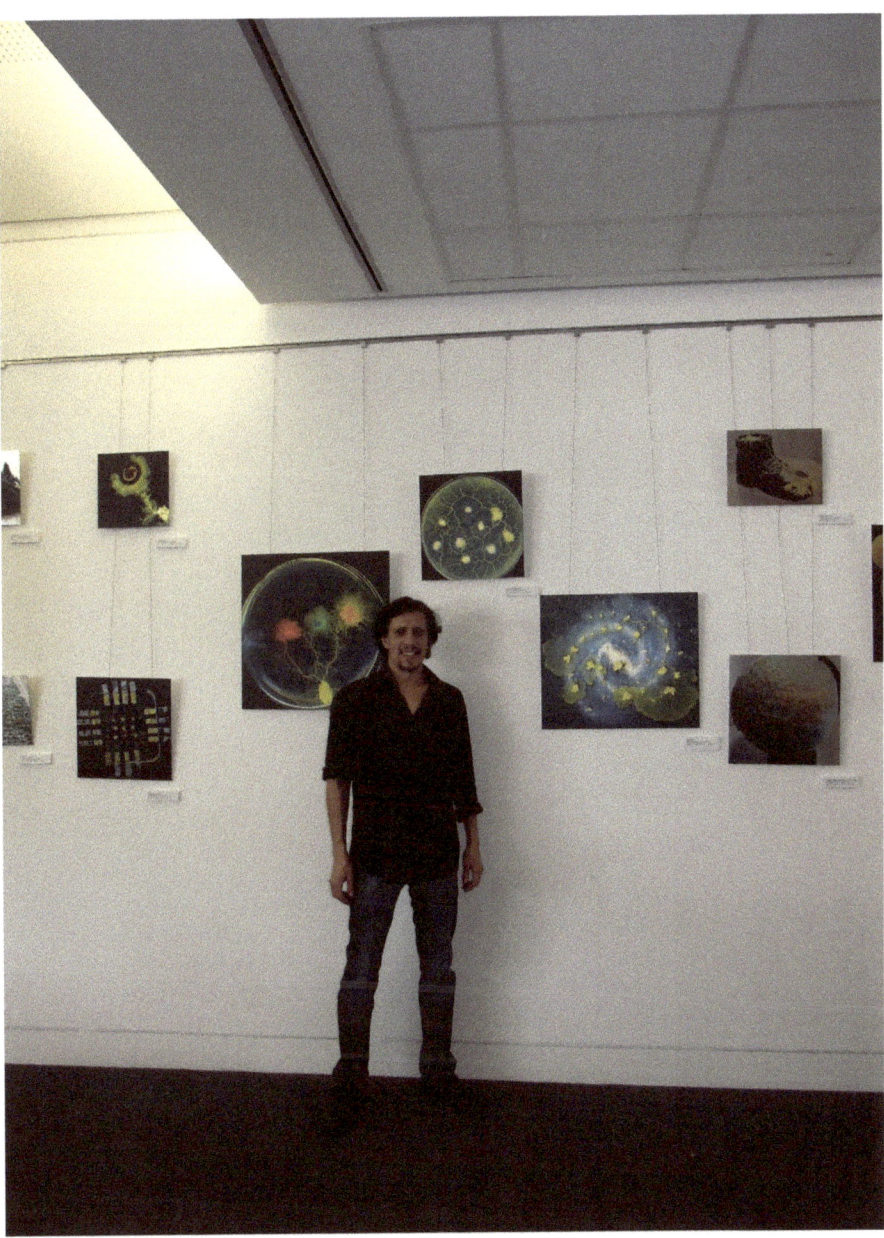

Andrew Adamatzky. Slime Mould Machines exhibition at Bristol Bright Night, 2014.

References

1. Adamatzky, A.: Computing in nonlinear media and automata collectives. Taylor and Francis (2001)
2. Adamatzky, A.: Developing proximity graphs by physarum polycephalum? Parallel Processing Letters **19**(01), 105–127 (2009)
3. Adamatzky, A.: Hot ice computer. Physics Letters A **374**(2), 264–271 (2009)
4. Adamatzky, A.: Manipulating substances with *Physarum polycephalum*. Materials Science and Engineering: C **30**(8), 1211–1220 (2010)
5. Adamatzky, A.: Physarum machines: Computers from slime mould. World Scientific (2010)
6. Adamatzky, A.: On attraction of slime mould physarum polycephalum to plants with sedative properties. Nature Proc (2011)
7. Adamatzky, A.: Skeletonization by crystallization. Physics Letters A **375**(3), 505–510 (2011)
8. Adamatzky, A.: Bioevaluation of World Transport Networks. World Scientific (2012)
9. Adamatzky, A.: Reaction-diffusion Automata: Phenomenology, Localisations, Computation. Springer (2012)
10. Adamatzky, A.: Simulating strange attraction of acellular slime mould *Physarum polycephaum* to herbal tablets. Mathematical and Computer Modelling **55**(3), 884–900 (2012)
11. Adamatzky, A.: Slime mold solves maze in one pass, assisted by gradient of chemo-attractants. NanoBioscience, IEEE Transactions on **11**(2), 131–134 (2012)
12. Adamatzky, A.: The world's colonization and trade routes formation as imitated by slime mould. International Journal of Bifurcation and Chaos **22**(08) (2012)
13. Adamatzky, A.: Geometry induced delays of slime mould propagation. Biophysical Reviews and Letters **8**(01), 89–97 (2013)
14. Adamatzky, A.: Physarum machines for space missions. Acta Futura **6**, 53–67 (2013)
15. Adamatzky, A.: Slimeware: Engineering devices with slime mold. Artificial life **19**(3-4), 317–330 (2013)
16. Adamatzky, A.: Towards slime mould colour sensor: Recognition of colours by *Physarum polycephalum*. Organic electronics **14**(12), 3355–3361 (2013)
17. Adamatzky, A.: Tactile bristle sensors made with slime mold. Sensors Journal, IEEE **14**(2), 324–332 (2014)
18. Adamatzky, A.: Towards plant wires. Biosystems **122**, 1–6 (2014)
19. Adamatzky, A.: Would be nervous system made of slime mould. Artificial Life (2014)
20. Adamatzky, A., Costello, B.D.L., Asai, T.: Reaction-diffusion computers. Elsevier (2005)
21. Adamatzky, A., Erokhin, V., Grube, M., Schubert, T., Schumann, A.: Physarum chip project: Growing computers from slime mould. IJUC **8**(4), 319–323 (2012)
22. Adamatzky, A., de Lacy Costello, B., Shirakawa, T.: Universal computation with limited resources. International Journal of Bifurcation and Chaos **18**(08), 2373–2389 (2008)
23. Mayne, R., Adamatzky, A.: Towards slime mould electrical logic gates with optical coupling. CoRR **abs/1403.3973** (2014)

24. Stephenson, S.L., Stempen, H., Hall, I.: Myxomycetes: a handbook of slime molds. Timber Press Portland, Oregon (1994)
25. Whiting, J.G., de Lacy Costello, B.P., Adamatzky, A.: Sensory fusion in *Physarum polycephalum* and implementing multi-sensory functional computation. Biosystems **119**, 45–52 (2014)
26. Whiting, J.G., de Lacy Costello, B.P., Adamatzky, A.: Towards slime mould chemical sensor: Mapping chemical inputs onto electrical potential dynamics of *Physarum Polycephalum*. Sensors and Actuators B: Chemical **191**, 844–853 (2014)

Index

active zone, 40
adder, 49
analog circuit, 5
Autobahn, 4, 22

Balkans, 4, 30
binary adder, 5

Cartwheel galaxy, 21
chip, 5, 12
choice, 11
colonisation, 4, 32, 34, 38
concave hull, 39
crystallisation, 5

delay, 14
dendritic tree, 5
digital circuit, 5
disc, 17

electrical potential, 5
encapsulation, 10
excitation, 47
exploration, 31
eye, 24, 44

galaxy, 33
germination, 6
graph, 50

highway, 19, 28
hull, 5

innervation, 26, 27

Mars, 4, 38
maze, 7, 43
Milky way, 20

neuron, 5
non-linear substrate, 3
nutrient, 23, 41

optical pathway, 5
optimisation, 3
oscillation, 5

Physarum, 12
    machines, 3
planning, 35
plant root, 5
propagation, 42
proximity, 50

reasoning, 3
road, 16
    networks, 4
route, 35

self-repair, 5

Silk Road, the, 4, 32
slime, 13, 16, 18, 22, 29, 37
   mould, 3

valerian, 5, 36
Voronoi diagram, 25, 48

waves, 47
whiskers, 37
wire, 13, 18, 29
wrapping, 9

www.ingramcontent.com/pod-product-compliance
Lightning Source LLC
Chambersburg PA
CBHW040544220526
45473CB00016B/3014

FACES OF THE MIDDLE EAST

PHOTOGRAPHY BY HERMOINE MACURA
www.facesofthemiddleeast.com

Printed in the U.S.A.

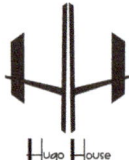

Hugo House Publishers, Ltd.
Denver, Colorado
Austin, Texas
www.HugoHousePublishers.com

FACES OF THE MIDDLE EAST
Copyright © 2015. Hermoine Macura. All Rights Reserved.

No part of this book may be reproduced or transmitted in any form or by any means, electronic or mechanical, including photocopying, recording, or by any information storage and retrieval system without written permission of the author or publisher.

ISBN: 978-1-936449-68-2

Library of Congress Control Number: 2015930752

Graphic Design & Image editing:
Sara Semlitsch, SARA JASMIN Design
Madelyn Aminikharrazi
COVA group, UAE.

Arabic calligraphy and Design by:
Eman Azab & Majid Al Yousef

Photography by Hermoine Macura
Images are taken with care and full knowledge of the people involved.